RHYTHMIC AND CONTRAPUNTAL STRUCTURES IN THE MUSIC OF ARTHUR HONEGGER

for Gene

Rhythmic and Contrapuntal Structures in the Music of Arthur Honegger

KEITH WATERS
University of Colorado at Boulder, USA

Routledge
Taylor & Francis Group
LONDON AND NEW YORK

First published 2002 by Ashgate Publishing

Reissued 2019 by Routledge
2 Park Square, Milton Park, Abingdon, Oxon OX14 4RN
52 Vanderbilt Avenue, New York, NY 10017

Routledge is an imprint of the Taylor & Francis Group, an informa business

A Library of Congress record exists under LC control number:

ISBN 13: 978-1-138-72869-1 (hbk)
ISBN 13: 978-1-315-18967-3 (ebk)

Contents

Preface

This book examines fundamental principles of musical organization in the music of Arthur Honegger. These principles are based upon systematic rhythmic and contrapuntal strategies, illuminated through an examination of Honegger's compositions and compositional sketches. While techniques of pitch organization are diverse in the compositions of Honegger, the strategies of rhythm and counterpoint provide an underlying source of compositional unity.

The initial chapter provides a brief historical summary of the composer's career and examines the significant compositional influences upon Honegger. A summary of the analytically-based secondary literature focuses upon issues of architectural symmetry and harmonic progression. A concluding review of Honegger's own prose writings examines a number of the composer's aesthetic positions.

The second and third chapters establish analytical methods for examining rhythmic and contrapuntal organization in Honegger's music. Chapter two discusses rhythmic and metric organization in several compositions, highlighting methods of rhythmic conflict, and shows how these same conflicts are expressed across different levels of rhythmic structure. The discussion attempts to extend and refine current theories of rhythmic (or "metrical") consonance and dissonance. Contrapuntal strategies are expressed through techniques of inversional symmetry, considered here in terms of contour symmetry, generic interval symmetry, and specific interval symmetry: this provides the focus of Chapter three. In addition to controlling outer-voice simultaneities, these and other symmetries, such as that of the interval cycle, govern longer-range linear spans.

Chapters four and five provide extended analyses of two orchestral works. *Mouvement symphonique no. 2 (Rugby)* establishes a series of correspondences based upon pitch material, rhythmic groupings, and rhythmic conflicts. Rhythmic conflicts are expressed at different levels of rhythmic and metric structure, and are stated in the domain of pitch. The analysis of the first two movements of *Symphonie pour cordes* is supplemented by the composer's sketches. The final chapter suggests avenues for future inquiry in the domain of pitch organization of the music of Honegger, in coordination with current analytical approaches for the tonally-centric music of Stravinsky, Bartók, and Hindemith.

* * *

Scholarship is in many ways a collaborative effort, and I would like to acknowledge the numerous people and institutions that have helped me in shaping my original dissertation and this book. Steve Bruns, Carlo Cabellero, Ralph Locke, and Robert Waters provided me with detailed and significant suggestions for the initial chapter. For the "Rhythmic Structures" chapter, Gretchen Horlacher, Harald Krebs, Steve Larson, Daphne Leong, and Debby Rohr each offered careful and thoughtful ideas. I also thank Kristin Taavola for her comments on Chapter three. My initial dissertation committee from the Eastman School of Music consisted of Daniel Harrison, Ralph Locke, and David Headlam, and was chaired by Robert Wason. Without their generous input this book would not be possible. Lezlie Botkin proofread the entire draft and offered meticulous and detailed editing suggestions. Jack Fredericksen patiently formatted many of the musical examples.

I would also like to acknowledge the support of the Eastman School of Music and its Department of Music Theory, which was chaired by Marie Rolf during my graduate studies there. The Paul Sacher Foundation in Basel, Switzerland provided me with a generous grant to study Arthur Honegger's compositional and contrapuntal sketches during the summer of 1994. The Foundation's Robert Piencikowski was unstinting in his help, and he offered many trenchant insights into twentieth-century French music and culture. Felix Meyer, the current director of the Sacher Foundation, suggested a number of fruitful avenues for inquiry. In addition, the *Association d'Arthur Honegger* and its director Annie Neuberger have provided an important focal point for research into Honegger's music. I would like to thank Editions Salabert and Carl Fischer for generously granting reprint permission for many of the musical examples used throughout the book. I would also like to thank Ashgate Publishing and Rachel Lynch for their interest in this project. Ruth Peters, Celia Hoare, and Sarah Charters of Ashgate cheerfully offered a good deal of editorial advice. Finally, this book owes a great deal to the unending support and sanity of Gene Hayworth.

Denver, Colorado

Chapter 1

Background

Introduction

> Honegger has vanished; with him our youth's initiation to contemporary music is obliterated. *Antigone*, *Horace Victorieux*, the *Symphonie pour cordes*: it is from there that we began a journey rich in surprises. Let us thank Honegger for having given us a taste for adventure.[1]

Pierre Boulez' elegy testifies to the influence that Arthur Honegger (1892-1955) exerted upon contemporary music of the first half of the twentieth century. Boulez identifies the music of Honegger as a signal point of departure for the path undertaken by a newer generation of composers. Other contributors to the 1956 "Hommage à Honegger" issue of *Melos*—Dallapiccola, Fortner, Henze, Messiaen, and others—discussed works most closely associated with the composer, works such as *Le Roi David*, *Mouvement symphonique (Pacific 231)*, and *Jeanne d'Arc au bûcher*, and they similarly acknowledged the influence of the recently deceased composer upon their own generation. Most singled out Honegger's ability to merge innovation and tradition, his contributions to both dramatic and symphonic music, and his stature as humanist and intellectual. The entries reflect Honegger's prestige as a composer at mid-century.

By the time of Honegger's death in 1955, he had written over 200 compositions. His output included five symphonies, over twenty ballet scores, chamber music, more than forty film scores, operas and operettas, music for voice, music for piano, as well as oratorios and other orchestral music. Although well-known as a member of the group of French composers *Les Six* in the early 1920s, Honegger composed most of his significant and best-known works after the group dissolved around the mid-1920s. His compositions incorporated elements of atonality, tonality, extended tonality, and polytonality, at times all within a single composition. Honegger's own credo combined accessibility with compositional rigor:

---

[1] "Honegger disparaît; avec lui s'efface l'initiation de notre jeunesse à la musique contemporaine ... *Antigone*, *Horace Victorieux*, la *Symphonie pour cordes*, c'est par là que nous avons commencé un périple fertile en surprises. Qu'Honegger soit remercié de nous avoir donné le goût de l'aventure." Pierre Boulez, Obituary from "Hommage à Honegger," *Melos* (January 1956): 4.

> My inclination and my effort have always been to write music which would be comprehensible to the great mass of listeners and at the same time sufficiently free of banality to interest genuine music lovers ... I have wanted to touch the two publics: the technicians and the multitude.[2]

Honegger's earliest works show the strong influence of other composers: the *Sonate pour violon et piano en ré mineur* (1912)[3] makes use of the cyclic form and the late-Romantic harmonic language of César Franck, while the *Prélude pour Aglavaine et Sélysette* (1917) and the *Rhapsodie* (1917) owe much to the harmonic language of Claude Debussy. Honegger's seven years of study at the Conservatoire in Paris (1911-1918)—which included classes with Charles-Marie Widor, Vincent d'Indy, and seven years of counterpoint study with André Gedalge—shifted the young composer's attention to chamber music, musical architecture (especially the use of sonata form), and counterpoint. Many of the works composed up to 1920 reflect these compositional preoccupations, including the String Quartet (1917), two Violin Sonatas (1918 and 1919), a Viola Sonata (1920), a Cello Sonata (1920), and a Sonatina for Two Violins (1920). In these compositions the harmonic language of late-Romanticism and Impressionism gives way to extended tonality and pitch centricity, and a number of the movements include fugal and imitative passages.

The works of the twenties veered into several different avenues. Some of the compositions were written in a lighter and more accessible idiom, at times reflecting the spirit of *Les Six*, such as the piano works *Sept pièces brèves* (1920), *Le Cahier romand* (1923), the *Concertino* for piano and orchestra (1924), and the Cello Concerto (1929). The incidental music for the play *Le Roi David* (1921) was turned into an oratorio two years later and launched Honegger's international career. In contrast, several works moved towards free atonality, including the orchestral work *Horace Victorieux* (1921) and the opera *Antigone* (1927). Honegger's best-known work, *Mouvement symphonique (Pacific 231)* (1923), was the first of three single-movement orchestral works that he termed *Mouvements symphoniques*. Rhythm plays a fundamental formal and structural role in these three works, which emerge from the tradition of the symphonic poem.

Owing to a decline in the number of commissions for concert works in the 1930s, Honegger turned towards more commercial endeavors. He composed twenty-four film scores, in addition to incidental music, radio plays, light cabaret songs, and several anthems for the leftist Popular Front political coalition. He

---

2 Arthur Honegger, *I Am a Composer*, trans. by Wilson O. Clough in collaboration with Allan Arthur Willman (New York: St. Martin's Press, 1966), 92, 94. Originally published as *Je suis compositeur* (Paris: Editions du Conquistador, 1951).

3 Unpublished, housed at the Paul Sacher Stiftung.

reached the second plateau of his career in the collaboration with poet Paul Claudel for the oratorio *Jeanne d'Arc au bûcher* (1935), which premiered in 1938.

In contrast to his facility in composing incidental music and film scores, Honegger acknowledged the difficulty and challenge of symphonic writing:

> As far as I am concerned, symphonic works give me much trouble: they demand an effort at sustained reflection. On the contrary, as soon as I can refer to a literary or visual pretext, my work becomes much easier.[4]

Although the first symphony was written in 1929, the last four were not written until after 1940. Some commentators consider these final four symphonies, created during the last fifteen years of the composer's life, to represent the summit of his output. One writer suggests, "It would be no exaggeration to say that Honegger, in his five symphonies, reveals the most substantial and human side of his inspiration—the very essence of his thought and art."[5] Significant works written in the final years of his life include the *Concerto da camera* (1948), the Fifth Symphony (*Di tre re*, 1950), and *Une cantata de Noël* (A Christmas Cantata, 1953).

Sources and Influences

Honegger stood at the confluence of various currents of musical tradition. As a music student he absorbed a wide array of musical styles that shaped his compositional approach. On the one hand, his earliest training at the Conservatory in Zürich between 1909-11 was steeped in the Germanic tradition. Geoffrey Spratt underlines the importance of these studies:

> The influence of this period, heavily biased towards a study of composition involving analysis of music by the Classical masters and the works of Wagner, R. Strauss and Reger, cannot be overstressed when the elements of Honegger's personal musical language are examined—at the very least they represent a framework that always distinguishes Honegger from the purely French school of composers.[6]

---

[4] Honegger, *Composer*, 80-81.

[5] Georges Favre, *Musiciens français contemporains*, vol. 2 (Paris: Durand, 1956), 77. Quoted and translated in Geoffrey Spratt, *The Music of Arthur Honegger* (Cork: Cork University Press, 1987), 368.

[6] Geoffrey Spratt, *The Music of Arthur Honegger* (Cork: Cork University Press, 1987), 2.

On the other hand, recent biographical evidence indicates a youthful and enthusiastic shift in musical allegiance after Honegger left the Conservatory in Zürich and immersed himself in Parisian musical culture upon entering the Conservatoire in Paris in 1911. In a letter to his parents in 1915, he emphasized this aesthetic change:

> ... my musical sympathies with the new French school are growing daily. There's no doubting that, for the first time, France is currently at the head of musical developments. I got to know and love Reger and Strauss while I was in Switzerland, and I continue to love them, but I realized that composers like Debussy, Dukas, d'Indy, Florent Schmitt, and others, were newer and more original, and above all contained more feeling than the modern Germans. Strauss's music, considered on its own and apart from its orchestral clothing that is its principal strength (and a strength of genius in that composer), is infinitely less rich in invention than the music of Debussy, for example, and often contains things that are extremely banal and old-fashioned. Reger's music is admirably written, but it often lacks real imagination or emotion, and for that reason constitutes an abuse of technical procedures.[7]

Honegger later acknowledged a détente between the French and German influences, noting that Debussy and Fauré provided a "useful counterbalance" to the classics and Wagner.[8] The influence of Debussy and Fauré is evident in Honegger's harmonic language, especially in the early vocal works that were written during Honegger's student years. These works, such as the *Trois Poèmes de Paul Fort* (1916), and the *Six Poèmes d'Apollinaire* (1915-1917), employ modality, as well as added seventh and ninth chords. Honegger's pronounced emphasis on musical architecture and large multi-movement works owed some debt to Vincent d'Indy, with whom Honegger had studied at the Conservatoire.[9]

In addition to his more immediate German and French predecessors, two earlier composers, Beethoven and Bach, were critical influences. Honegger wrote of his interest in and study of Beethoven sonatas as a youth in Le Havre, and some of his earliest compositional forays were written "quite naïvely in the

---

[7] Originally quoted in Harry Halbreich, *Arthur Honegger: Un musicien dans la cité des hommes* (Paris: Librairie Arthème Fayard, 1992), 36-7. Trans. by Roger Nichols in *Arthur Honegger* (Portland, Oregon: Amadeus Press, 1999), 32.

[8] Honegger, *Composer*, 91.

[9] Honegger studied conducting with d'Indy beginning in March 1916. He later described these classes with d'Indy as workshops in which the composition students could hear their works played and could improve on compositional or orchestrational problems. See Halbreich, *Honegger*, trans. Nichols, 35.

style of Beethoven."[10] Beethoven's influence is especially apparent in Honegger's reliance on traditional strategies for thematic development, manipulation, and juxtaposition. These strategies provide a consistent source of musical organization throughout Honegger's *oeuvre*.

Writing in 1925, Honegger discussed the more pervasive influence of Bach:

> But the chief of them, he whom I never cease to take as a model, is old Bach. In him I find all that I wish ... One might just as easily find Bach at the origins of all my works.[11]

Bach's influence was manifold. A Protestant, Honegger was attracted to the genre of oratorio and to biblical subjects, reflected particularly in his works *Le Roi David* (1921) and *Judith* (1925). Honegger allied himself with Bach's ethos of composer as religious craftsman. In addition, Honegger overtly modeled his own contrapuntal orientation upon Bach's.[12]

The influences of Beethoven and Bach consistently supplied Honegger with traditional methods for musical organization: in the case of Beethoven, the commitment to thematic developmental techniques, and in the case of Bach, the use of counterpoint and contrapuntal devices of fugue, canon, stretto, and imitation. Like other twentieth-century neoclassical composers, Honegger often employed these methods independent of common-practice harmonic constraints, instead using a focal pitch-class to articulate a tonal center at formal arrival points.

Finally, three twentieth-century figures formed important influences: Schoenberg, Stravinsky, and Bartók. Halbreich's recent biography discusses the attraction of Arnold Schoenberg for the Conservatoire students Honegger and Milhaud. Serving in the military in Switzerland between 1915 and 1916, Honegger was able to obtain musical scores of Schoenberg unavailable in wartime Paris, including the song cycle *The Book of the Hanging Gardens* (op. 15, 1908).[13] Honegger also attempted to translate Schoenberg's *Harmonielehre*, a

---

10 Arthur Honegger, *Nachklang, Schriften, Dokumente*, ed. by Willi Reich (Zürich: Die Arche, 1957), 9. "All dies war ganz naiv im Stile Beethovens geschrieben, denn in den Klaviersonaten dieses Meisters hatte ich, ehe ich noch Unterricht erhielt, die Grundgesetze der Harmonie entdeckt und verstanden." Honegger is here referring to his unpublished six sonatas for violin and piano, written in 1908 at the age of sixteen.

11 Quoted and translated in Kurt von Fischer, "Arthur Honegger," included in *Essays in Musicology*, trans. by Carl Skoggard and ed. by Tamara S. Evans (New York: The Graduate School and University Center, City University of New York, 1989), 154-155. Originally appeared in "Opinions d'Arthur Honegger," *Dissonances* II (1925): 84-7.

12 See the quote p. 3 above. Honegger's contrapuntal orientation and his contrapuntal studies at the Conservatoire for the focus of Chapter three.

13 Halbreich, *Honegger*, trans. Nichols, 33.

text that he described as "an extraordinary work and indispensable for the modern composer."[14] Honegger seems to have studied the implications of Schoenberg's treatise in quite specific ways. A chord from Honegger's sketchbook is illustrative of this connection. Example 1.1a shows this chord from the sketchbook, consisting of all twelve pitch classes derived from the vertical combination of three diminished-seventh chords. Example 1.1b is from Figure 339g of Schoenberg's *Harmonielehre*, taken from the chapter entitled "Chords with six or more notes,"[15] similarly showing a twelve-note chord distributed through the combination of three diminished-seventh chords.

Example 1.1a and b: 12-note Chord in Honegger and Schoenberg

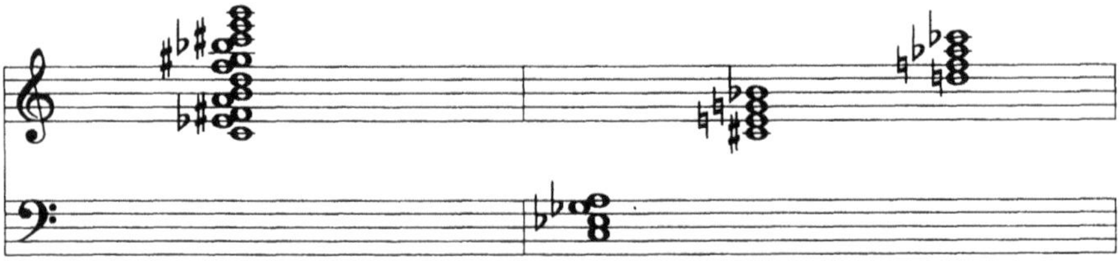

1a. Honegger: from sketchbook   1b. Schoenberg: *Harmonielehre*

The use of chromatic completion in a number of Honegger's works also suggests another link with Schoenberg's treatise as well as with Schoenberg's pre-serial atonal compositions.[16] In one analysis, Maillard and Nahoum illustrate Honegger's strategy of using all twelve pitch classes in both the horizontal and the vertical dimension, and they note the overt—if infrequent—technique of harmonizing a melodic cell with the residual pitches of the chromatic scale.[17]

Example 1.2, from the beginning of the second movement of Honegger's

---

[14] Ibid., 35.

[15] Honegger's sketch taken from microfilm 185: 325 at the Paul Sacher Foundation. It is not clear what, if any, composition this was intended for. Proximate sketch material suggests entry between 1921-1922. Schoenberg's chord taken from Arnold Schoenberg, *Theory of Harmony*, trans. Roy E. Carter (Berkeley and Los Angeles: University of California Press, 1978), 407.

[16] For the music of Schoenberg, Ethan Haimo distinguishes between chromatic completion, in which a pitch class may return before all twelve have sounded, and an aggregate, in which no pitch class may so recur. Ethan Haimo, *Schoenberg's Serial Odyssey: The Evolution of his Twelve-Tone Method, 1914-1928* (Oxford: Clarendon Press, 1990), 44, 183.

[17] Jean Maillard and Jacques Nahoum, *Les symphonies d'Arthur Honegger* (Paris: Alphonse Leduc, 1974), 60.

1920 vocal composition *Trois Fragments extraits de "Les Pâques à New York" de Blaise Cendrars*, demonstrates the technique of keeping all twelve pitch classes in circulation. Figure 1.1, below Example 1.2, indicates each of the pitch classes used in each measure. Figure 1.1 shows that by m. 2, any pitch class not stated in a given measure appears in the following measure.

Example 1.2: Mm. 1-9 from *Les Pâques à New York*, ii

Example 1.2 (Continued)

Figure 1.1: Chromatic completion in mvmt. 2 of *Les Pâques à New York*, mm. 1-9 (C = 0, C# = 1 ... Bb = A, B = B)

M. 1:	0								8	9	A	
2:	0	1	2	3			6	7	8	9	A	B
3:	0			3	4	5	6	7	8	9	A	
4:		1	2	3	4			7	8	9		B
5:	0		2	3	4	5	6		8	9	A	
6:	0	1	2	3	4		6	7	8	9	A	B
7:	0		2	3	4	5	6	7	8	9	A	
8:	0	1	2	3	4	5	6		8	9		B
9:	0	1		3	4		6	7		9	A	B

In addition to Schoenberg, Stravinsky plays a significant role in the understanding of Honegger's compositional style, particularly in regard to Honegger's forceful rhythmic language, the use of bitonality, and pitch centricity.

Honegger identified a third pivotal twentieth-century precursor:

> The writers on music who deal with this post-Debussy period generally agree in recognizing Schoenberg and Stravinsky as those responsible for the reaction that it has stimulated. Certain writers would even add Erik Satie. To my thinking, it is much more Bartók who, with Schoenberg and Stravinsky, is the true representative of this generation's musical revolution. Less direct, less sparkling than Stravinsky, less dogmatic than Schoenberg, he is, of the three, perhaps

the most profoundly musical, and the one whose development has taken place by means of the most continual and best organized effort.[18]

Musical correspondences between Bartók and Honegger have not been developed significantly in biographical and analytical literature, but extant observations are suggestive. First, Honegger returned to Paris in 1916 with scores of the music of Bartók, in addition to those of Schoenberg. (See footnote 13 above.) Second, Honegger's preoccupation with formal symmetry through reversed recapitulation and *Bogenform* closely mirrors the formal techniques of Bartók. Third, one of Geoffrey Spratt's primary analytical observations is Honegger's penchant for harmonic motion by thirds, especially by minor thirds. This aligns to a certain degree with Ernő Lendvai's axis-system construct for the compositions of Bartók.[19] Moreover, Lendvai and others have called attention to Bartók's use of the golden-section proportion to structure a number of his compositions. Honegger uses the golden-section proportion also in several works, beginning as early as 1917 in his *Le Chant de Nigamon*, and appearing as late as 1950 in one of his final compositions, the third movement from the Symphony no. 5.[20]

Honegger and *Les Six*

Honegger is perhaps best known for his membership in the group of French composers *Les Six* in the early 1920s. Together with Honegger, *Les Six* consisted of Darius Milhaud, Francis Poulenc, Georges Auric, Germaine Tailleferre, and Louis Durey. The six composers, in loose collaboration with Jean Cocteau and Erik Satie, had already had their works played jointly on concerts since 1918, often billing themselves as *Les Nouveaux jeunes*. The name *Les Six* was derived from two articles by Henri Collet that appeared in the journal *Comoedia* in January 1920 entitled "Un ouvrage de Rimsky et un ouvrage de ... Cocteau: les

---

[18] Quoted and translated in von Fischer, "Arthur Honegger," 153. This quote originally appeared in Honegger's introduction written for Serge Moreux, *Belá Bartók, sa vie, ses oeuvres, son langage* (Paris: Richard Massé, 1949), 7-9.

[19] See especially Ernő Lendvai, *Béla Bartók: An Analysis of His Music* (London: Kahn and Averill, 1971); and *The Workshop of Bartók and Kodály* (Budapest: Editio Musica, 1983). Geoffrey Spratt's observes Honegger's harmonic practices in his book: see Spratt, *Arthur Honegger*. Some of the problems in determining harmonic areas are discussed later in the chapter.

[20] See Halbreich, *Honegger*, trans. Nichols, 336 and 347 for a discussion. See also Chapter five of this work, where I discuss the use of the golden-section proportion in the second movement of the *Symphonie pour cordes* (Symphony no. 2). In addition to Bartók, another potential source for the use of this proportion may have come about through Honegger's association with poet Guillaume Apollinaire, who was affiliated with the group of cubist painters known as the "Section d'or" in 1912.

Cinq Russes, les Six Français" and "Les Six Français." The resultant publicity from Collet's articles, fueled by Jean Cocteau's vigorous pamphleteering, helped establish a public perception of the six composers as standard-bearers for music that was distinctly French, unpretentiously rooted in popular music sources of the music hall, circus, and cabaret, and characterized by flippancy, levity, deliberate banality, and mordant Gallic wit.[21]

Yet, as the composers themselves insisted, the notion of a monolithic group aesthetic was a vast oversimplification, and their relationship was based upon mutual friendship rather than upon a shared set of musical values. In particular, Honegger's inclusion in *Les Six* is frequently offered as evidence of the arbitrary nature of their association, given Honegger's interest in larger musical structures and his affinity for German composers. Less than a year after Collet's articles had appeared in January 1920, Honegger distanced himself from the perception of *Les Six* largely molded by Cocteau. Posing instead as a traditionalist, formalist, and a champion of absolute music, he declared:

> I do not cultivate an admiration for the music of the fairgrounds and music hall, but on the contrary for that of chamber and symphonic music at its most grave and austere ... I attach great importance to musical architecture which I would never like to see sacrificed for literary or pictorial reasons. I also have a perhaps exaggerated tendency to return to polyphonic complexity. My great model is J.S. Bach.[22]

On the surface, Honegger's comments seem to strongly repudiate several positions associated with *Les Six*, and indicate an aesthetic indebted more to German than to French musical values. Some writers have suggested that the values espoused in these comments drastically set him apart from the other composers of the group and from Cocteau and Satie.[23] Certainly, Honegger's notion of chamber and symphonic music "at its most grave and austere" contradicts the spirit of levity and frivolity that characterizes many of the

---

[21] Cocteau's published his pamphlet *Le Coq et l'Arelequin* in 1918, and four issues of *Le Coq* were released between May-November 1920.

[22] "... je ne cultive pas l'admiration de la Foire et du Music-hall, mais au contraire celle de la musique de chambre et de la musique symphonique dans ce qu'elle a de plus grave et austère... J'attache aussi une grande importance à l'architecture musicale que je ne voudrais jamais voir sacrifiée à des raisons d'ordre littéraire ou pictural. J'ai aussi une tendance peut-être exagérée à rechercher la complexité polyphonique. Mon grand modèle est Jean-Sébastien Bach." Arthur Honegger, *Ecrits*, textes réunis et annotés par Huguette Calmel (Paris: Librairie Honoré Champion, 1992), 34. Portions of this letter to journalist Paul Landormy originally appeared in "MM Honegger et Georges Auric," *La Victoire* (September 28, 1920): 2.

[23] Regarding Honegger's 1920 statements quoted above, Halbreich writes, "There is in fact practically nothing in that paragraph, except an admiration for J.S. Bach, to which the other members of *Les Six*, Cocteau, or Satie, would have subscribed." See Halbreich, *Honegger*, trans. Nichols, 66.

representative works of *Les Six*, especially those by Poulenc, Milhaud, and Auric. In addition, Honegger and Satie were dismissive of each other's music, a hostility sparked in part by the friction between Honegger's musical *gravitas* and Satie's calculated banality. Moreover, Honegger's love of the music of Wagner contradicted the anti-Wagnerian maxims churned out by Cocteau. Honegger and Cocteau carried out an epistolary war in the press in 1922 with Honegger adopting the stance of prickly traditionalist, writing:

> ... [I]t seems to me indispensable that, to progress, we must be solidly linked with what has preceded us. We must not break the tie with musical tradition. A branch separated from the trunk quickly dies ... Economy of means seems to me more difficult, but also more useful, than a too headstrong audacity. There is no profit in smashing the door which you might open.[24]

However, while it is tempting to overdramatize Honegger's outsider status in relation to *Les Six*, it is important to take a nuanced view of the group and of Honegger's relationship to it. This relationship was a complicated one and shifted over time. While the evidence reveals some fractious divisions, it also shows several striking musical correspondences between Honegger and the rest of *Les Six*. First, Honegger's lack of interest in drawing from popular music sources—"music of the fairgrounds and the music hall"—was shared both by Tailleferre and by Durey. Only Poulenc, Auric, and Milhaud, in contrast, made explicit use of popular music sources.[25]

Second, in emphasizing musical architecture, chamber, and symphonic music as early as 1920, Honegger even anticipates the work of the Milhaud-Auric-Poulenc wing of *Les Six*. By 1924, Milhaud, Auric, and Poulenc had each adopted classical forms and models, *replacing* their use of popular music sources.[26] Finally, Honegger's quotation above accords an enhanced role for polyphony in his own music. Although individual practice differed, this linear orientation was largely shared by all of the group.

These last two components, the regeneration of classical forms and the linear orientation, broadly constructed a shared neoclassical aesthetic among *Les Six*. A clarity of form and line was to provide an alternative to the harmonic and formal vagueness of Impressionism. As Milhaud maintained, "What musicians asked for now was a clearer, sturdier, more precise type of art that should not yet

---

[24] Honegger, *Composer*, 93-4. The quote originally appeared in Honegger's article "Petit historique nécessaire," *Le Courrier musical* (February 1, 1922).

[25] For a more detailed discussion of the role of popular music sources in *Les Six*, see Nancy Perloff, *Art and the Everyday: Popular Entertainment and the Circle of Erik Satie* (Oxford: Clarendon Press, 1991).

[26] Ibid.

have lost its qualities of human sympathy and sensitivity ... After all the vapors of impressionism, would not this simple, clear art renewing the tradition of Mozart and Scarlatti represent the next phase in the development of our music?"[27]

It is easy to identify the undeniable and profound musical differences separating Honegger from the other members of the group, as well as from Satie and Cocteau. Yet the common musical elements linking Honegger and the other members of *Les Six* suggest a more complex musical relationship than that provided by the image of Honegger as sole dissident. Honegger did participate in the two collaborative efforts of the group, supplying the piano composition *Sarabande* for the *Album des Six* in 1920, and the following year contributing *La Noce massacrée* for Cocteau's stage work *Les Mariés de la tour Eiffel*. In addition, some of Honegger's compositions written during the 1920s indeed reflected the lighter French tradition attributed to the group, such as the piano works *Sept pièces brèves* (1920), *Le Cahier romand* (1923), the vocal work *Six Poésies de Jean Cocteau* (1923), as well as the *Concertino* for piano and orchestra (1924) and the Cello Concerto (1929). In their lucid diatonicism, simple four-bar phrasing, use of humor, and jazz references, these compositions share characteristics with other representative works of *Les Six*.

Literature on Honegger: Analytical Orientations

Literature on Arthur Honegger began accumulating as early as the mid-twenties,[28] and Honegger's own writings appeared in print beginning in 1920. The earliest studies gravitated to biography, and subsequent works attempted to classify the composer's work through chronology and genre.[29] Since Honegger's death in 1955, analytically-based studies have appeared, and these works attempt to address systematically the musical content of the compositions. Several of these studies establish larger categories based on the composer's compositional aesthetics and musical techniques;[30] others, especially since the mid-1970s,

---

[27] Darius Milhaud, *Notes Without Music*, trans. Donald Evans, ed. Rollo Myers (New York: Alfted A. Knopf, 1953), 95. This aesthetic also paralleled that of the *neue Sachlichkeit* of Paul Hindemith. See David Neumeyer, *The Music of Paul Hindemith* (New Haven and London: Yale University Press, 1986), 2.

[28] Roland-Manuel, *Arthur Honegger* (Paris: Editions Senart, 1925); and André George, *Arthur Honegger* (Paris: Claude Aveline, 1926).

[29] José Bruyr, *Arthur Honegger et son oeuvre* (Paris: Editions Corrêa, 1947); Marcel Delannoy, *Arthur Honegger* (Editions Horay, 1953); Jacques Feschotte, *Arthur Honegger* (Paris: Editions Seghers, 1966); Willy Tappolet, *Arthur Honegger* (Zürich: Hug, 1933; updated and reprinted, Zürich: Hug, 1954; translated into French, Neuchâtel: La Baconnière, 1957).

[30] Marcel Landowski, with the collaboration of de Marie Françoise Christout, *Arthur*

provide more thorough and rigorous technical discussion of the music, often narrowing the scope of inquiry to one facet of Honegger's *oeuvre*, such as the symphonies or songs.[31]

A significant analytical problem arises from the volume and diversity of Honegger's works. The output of more than two hundred compositions is characterized by an eclecticism of genre, and techniques of pitch organization within the compositions are equally eclectic. This diversity strongly challenges the possibility of a single analytical approach, and has led many analysts and biographers to rely on general description and stylistic classification.

In addition, in light of the *absence* of a systematic and singular method for pitch organization, a significant portion of the analytical literature focuses primarily on the thematic and formal organization of Honegger's works. This approach may be expected, given the composer's cultivation of large-scale musical works; moreover, his prose writings suggest that his notion of musical form was—to a large extent—thematically driven. Yet, in the literature, formal and thematic analysis alone has encouraged a chronological narrative of events within a composition, serving more as description than analysis.

However, several recurring issues come to light in the analytical literature. Since these issues illustrate deeper aspects of musical structure and organization and point to underlying compositional and aesthetic features within Honegger's works, they are more analytically revealing than taxonomic or thematic and formal descriptions. It is useful to isolate several of these analytical

---

Honegger (Paris: Editions de Seuil, 1957); Jean Matter, *Honegger, ou la Quête de Joie* (Lausanne: Foetisch Frères, 1956); Pierre Meylan, *Arthur Honegger: Humanitäre Botschaft der Musik* (Frauenfeld und Stuttgart: Verlag Huber, 1970); Elena Sysoeva, *Simfonii A. Honeggera* (Moscow: Muzyka, 1975); Kurt von Fischer, *Arthur Honegger*, Neujahrsblatt der Allgemeinen Musikgesellschaft (Zürich: Hug, 1978), translated and reprinted in Fischer's *Essays in Musicology*, Tamara Evans, ed., trans. Carl Skoggard (New York: Graduate School and University Center, City University of New York, 1989), 145-177.

[31] Halbreich, *Honegger*; Halbreich, *L'oeuvre d'Arthur Honegger* (Paris: Honoré Champion Editeur, 1994); Spratt, *Arthur Honegger*. Works which treat the symphonies alone include Jean Maillard and Jacques Nahoum, *Les symphonies d'Arthur Honegger* (Paris: Alphonse Leduc, 1974); Daniel Quienton Parker III, "The Symphonies of Arthur Honegger: An Examination of Stylistic Factors and Form" (Ph.D. diss., University of Kentucky, 1990); Mary Louise Shadel, "Thematic Materials and Developmental Procedures in the Slow Movements of the Five Symphonies of Arthur Honegger" (Master's Thesis, Georgia State University, 1982); Laurent Fichet, "Problèmes de langage dans l'oeuvre symphonique d'Arthur Honegger," contained in *Honegger-Milhaud: Musique et Esthétique*, ed. Manfred Kelkel (Paris: Librairie J. Vrin, 1994). Fichet also engages in numerological speculation for portions of the Fifth Symphony in "Symbolisme caché chez Honegger," *Bulletin de l'Association Arthur Honegger* No. 2 (October 1995): 7-11. Two works examining the songs of Honegger are Cindy Neumayer Dewey, "A Performer's Guide to the Songs of Arthur Honegger" (DMA Thesis, The Louisiana State University and Agricultural and Mechanical College, 1994); and Leslie Ann Goldberg, "An Analysis of Harmonic Language in the Songs of Arthur Honegger" (Mus.A.D. Thesis, Boston University, 1991).

threads, presented below under the headings of symmetry and harmonic progression.

Symmetry

One of the earliest discussions of formal symmetry appears in Willy Tappolet's study of Honegger. Tappolet points out the importance of formal symmetry to the composer in a number of works, including the ABA construction and durational[32] symmetry of *Chant de Joie* and the formal and durational symmetry of *Monopartita*.[33] Tappolet's diagram of the second movement of the *Première Symphonie*, provided in Example 1.3, graphically indicates the principle of what he calls an "absolute thematic symmetry." The diagram shows the archform symmetry of the eight sections of the movement as well as durational symmetry, in which musical material occupies nearly the same number of measures at its recurrence. Of the eight sections (*périodes*), the first and eighth are paired, the second and seventh, the third and sixth, and the fourth and fifth.

Example 1.3: Symmetry in *Première Symphonie* (from Tappolet, p. 235)

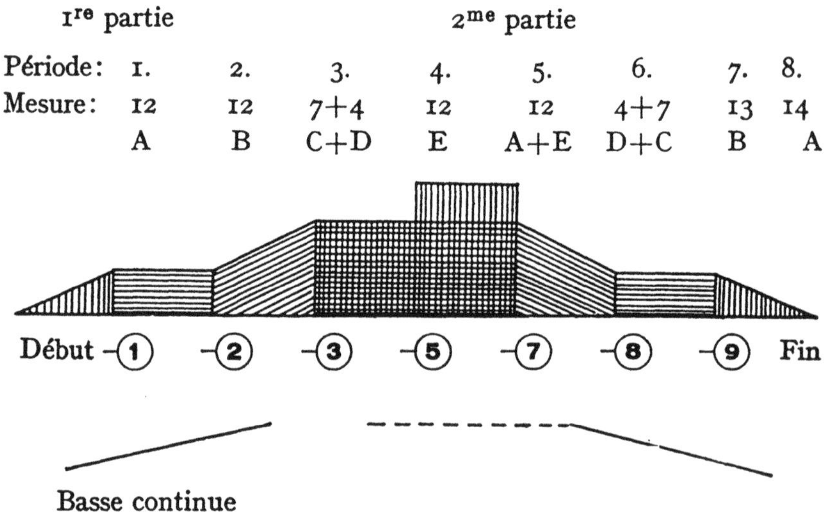

32 "Durational" symmetry in this context will refer to equal (or nearly-equal) lengths for parallel sections within a composition, usually determined by the number of measures.

33 Tappolet, *Arthur Honegger*. Tappolet discusses *Chant de Joie* on p. 88 and *Monopartita* on pp. 262-3.

In addition, the diagram points out the beginning and ending symmetry provided by the chromatic bass ascent of mm. 1-24, and the chromatic bass descent of the final measures of mm. 84-97. The diagram is somewhat misleading, however, in that the thirteen measures of Tappolet's "*période 7*," corresponding to Rehearsal number 8 below the schematic drawing, in fact involve a rising chromatic bass between Rehearsal 8 and 9, in contrast to the descending direction of the "basse continue" of the diagram.

Composer Marcel Landowski likewise addresses the issue of compositional symmetry in Honegger's works in a short biographical and analytical work that appeared two years after Honegger's death.[34] Landowski points out the archform symmetry of three of the early chamber works, including the second movement of the first String Quartet (1917), written in a nine-part form of ABCDEDCBA; as well as the ABCBA form of both the first Sonata for Violin and Piano (1917) and the Sonata for Viola and Piano (1920).

The Swiss musicologist Kurt von Fischer places the use of formal symmetry within a larger framework, a framework that identifies the composer's global aesthetic principles. He states, "though it was characteristic of Honegger to cultivate many different genres, his music exhibits certain constants that may be understood as the musical-material manifestation of basic aesthetic principles."[35] In his discussion of *Monopartita*, Fischer indicates the large-scale symmetry of the four sections of this one-movement work: within the ABBA form, the two Largo A sections are 21 and 14 measures, respectively; and the B sections are both 122 measures. The analyst goes on to note, however, the *lack* of symmetry within each of the B sections. The three subsections of the first B section (Vivace-Toccata-Adagio) are constructed of 71, 19, and 32 measures respectively; those of the second B section (Vivace-Adagio-Toccata) are 84, 32, and 6 measures. For Fischer, within the larger symmetries, these "rearrangements and irregularities" create their effect through the manipulation of musical time.[36]

Harmonic Progression

Fischer also undertakes a brief harmonic analysis of the final three measures of the fughetta from *Prélude, Arioso, Fughette (sur le nom de BACH)*. His terminology and approach is conditioned by the functional categories of Hugo Riemann. In addition, Fischer implicitly engages the notion of an extended tonality.

---

34 Landowski, *Arthur Honegger*.

35 von Fischer, *Essays in Musicology*, 160.

36 Ibid., 166.

> The basis of Honegger's harmonic practice always lies in tonic-dominant connections and the tensional relationships of dissonance as opposed to consonance; both of these phenomena are to be taken in a broadened sense, however.[37]

Fischer asserts, categorically, that Honegger's tonality is consistently rooted in the notion of chordal and harmonic function. His following analytical commentary augments this notion of tonality through the ideas of an added-chord harmonic language, as well as through harmonic substitution.

Example 1.4a contains the passage. For convenience, the chords are numbered here one through four. Fischer states that the first chord is the enharmonic equivalent of the dominant of the final B major triad, with an added second (Ab^4), and with the dominant seventh replaced by a major seventh (F^5). Continuing, Fischer maintains that the second chord is to be understood as a minor subdominant (i.e., E minor), with the added tones of C#, D#, and A. The penultimate chord, combining Bb minor with C minor, substitutes for the F# dominant. For Fischer, the substitute chords provide degrees of tension, which are increased and decreased through the chordal progressions, serving the "emotional side of Honegger's music."

However, Fischer's vertically oriented analysis neglects the excerpt's linear component. The relationship of the outer voices, especially, suggests a contrapuntal derivation typical of Honegger's voice-leading practices. As the analytical reduction of Example 1.4b shows, between the first and second chord the outer-voice contrary motion is literal: the Gb^2-G of the bass is mirrored by the Bb^5-A of the upper part. Moreover, the contrary motion is extended to the outer voices for both the right hand and left hand: the left-hand tenths ascend by half step (Gb/Bb—G/B) at the same time that the right-hand ninths descend by half step (Ab/Bb—G/A).

In the second measure, the outer-voice relationship between Chords 2 and 3 is likewise based upon contrary motion. Here, however, the relationship is not intervalically exact: the ascending minor third A^5-C of the soprano is stated against the descending whole step of G^2-F in the bass, preserving the B-A-C-H motive in the soprano.

Fischer makes an intriguing analytical claim when examining the final two chords of the progression (Chords 3-4). He suggests that the simultaneous statement of Bb minor and C minor in the penultimate chord "encircles" the ultimate B major resolution, and that this Chord 3 substitutes for an F# dominant chord. This is a departure from his earlier analytical labels. In Chords 1 and 2, Fischer was able to justify the use of traditional functional labels through added pitches combined with a foundational triad (Gb/F# = dominant in the case of

---

[37] Ibid., 162.

Example 1.4a: Final three measures of fughetta from
Prélude, Arioso, Fughette (sur le nom de BACH)

Example 1.4b: Outer-voice inversional symmetry

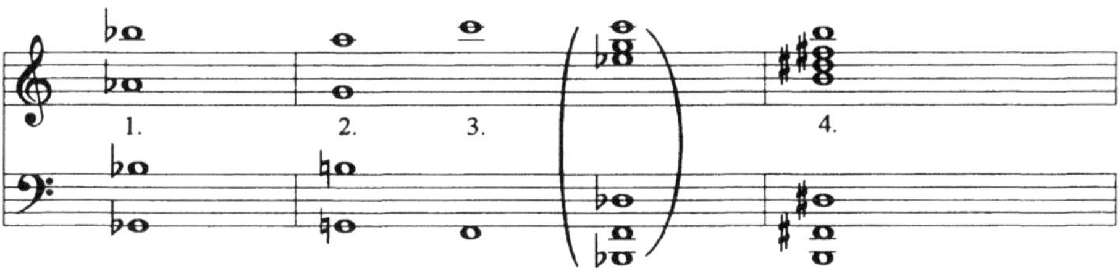

Chord 1, E minor = subdominant in the case of Chord 2). However, in Chord 3 he is unable to identify a traditional functional dominant and offers instead a polychordal dominant substitute, in which Bb minor in the left hand is combined with C minor in the right hand.

His claim that this Bb minor/C minor dominant substitute encircles the final B major triad has interesting analytical implications. It suggests that a tonal center may be preceded by and articulated by a harmony whose root is a half step above, below or—in this case—both above and below the ultimate harmonic destination. In addition, Fischer's terminology ("encircles") implies a linear derivation for the progression from Chord 3 to Chord 4—a potential contrary motion move in which the lower Bb minor chord resolves upward by half step and the upper C minor chord resolves down by half step, both converging on B major. However, the progression engenders contrary motion only abstractly. Literal contrary motion in the outer voices would only arise from the use of a root position Bb minor triad in the left hand at Chord 3, but this harmony is instead stated in inversion. Example 1.4b models this hypothetical contrary motion progression in the parentheses following Chord 3.

Fischer's attempts to invoke and expand upon the functional categories of tonic, subdominant, and dominant reveal some of the problems in analyzing harmonic progression in Honegger's music. In this instance, the harmonic veneer is primarily triadic: the final harmony (Chord 4) of the composition is an unadorned B major triad, and it is possible to deduce triadic foundations from Chords 1, 2, and 3. However, the harmonic progression—the chord-to-chord succession—makes conventional harmonic interpretation slippery and elusive. Moreover, Fischer's harmonic orientation underemphasizes the linear component operating in Honegger's progression. Honegger's study and cultivation of counterpoint (discussed in Chapter 3) suggests that a synthesis of both vertical and horizontal elements is necessary when taking into account harmonic progression.

Another analyst who has grappled with harmonic and tonal progression is Geoffrey Spratt, whose book *The Music of Arthur Honegger* appeared in 1987. [38] Spratt goes further than Fischer by analyzing a number of complete compositions. His formal diagrams include the principal tonalities for compositional sections. However, Spratt does not make explicit his criteria for tonal determination. Given the tonal ambiguity of many of the works (or portions of the works), it is unclear how these decisions are made, and this potentially undermines analytical and stylistic observations that depend upon these key relationships.[39]

For example, Spratt calls attention to harmonic movement by thirds throughout many of Honegger's compositions. [40] In his discussion of Mouvement *Symphonique #2, Rugby*, this observation leads him to conclude that "typical of Honegger's fastidious attention to aesthetic balance and proportions, however, is firstly the return in between the pillars of C# major (Reh. 8-11) to the tertiary related tonal centers (G major, B major and G minor) which had dominated the first episode ..."[41] The extended passage warrants further investigation. Example 1.5a contains Honegger's piano reduction of the orchestral score; Example 1.5b contains a harmonic reduction of those measures.

Although Spratt does not give the location of these tonal centers of G

---

[38] Spratt, *Arthur Honegger*.

[39] For example, Spratt makes general but sometimes unsubstantiated assertions regarding compositional unity based upon the relationship of key scheme to melodic intervals within a composition. In his discussion of the String Quartet no. 1 and the Violin Sonata no. 1, he states: "...the pronounced system of tonal organization involving tertiary relationships is exploited here to most effective ends. The tertiary intervals dominate the melodic construction too, so it is hardly surprising that the music exhibits extreme qualities of organic cohesion." Ibid., 8.

[40] See, for example, the discussions of the String Quartet no. 1 and the Violin Sonata no. 1 (p. 8); *Le Chant de Nigamon* (p. 18); *Amphion* (p. 173); Symphony no. 2 (p. 414) in ibid.

[41] Ibid., 157.

major, B major, and G minor in R. 8-11, he is most likely referring to the G major chord at two measures before R. 9 (labeled "1" in Examples 1.5a and b), and the B major chord in first inversion at R. 9 (labeled "2" in the examples). A G minor triad appears twice, once in the lower voices four measures after R. 9 (labeled "3"), and again four measures after R. 10 (labeled "4").

The first five measures of the example are diatonic in the key of Db major, and the added-chord harmonies over the bass pitches of Db, Eb, and F assert tonic, supertonic, and first inversion tonic chords over the stepwise bass. Example 1.5b shows how the ii7 chords may be interpreted in a contrapuntal fashion, operating first as a passing chord between I and I6 (one measure after R. 8), and as a neighboring harmony between two statements of I6 (three measures after R. 8). The diatonicism shifts upward a half-step from Db to D for the next three measures, and the bass pitches F#, G, and A on the downbeats assert a first inversion tonic, subdominant, and dominant within the key of D major. Thus, given the harmonic context, this G major harmony (six measures after R. 8, labeled as 1) does not operate as a tonal center, but is subordinate to the local temporary tonality of D major.

Spratt's second tonal center of B major presumably corresponds to the arrival on the first inversion B major triad at R. 9 (labeled as 2). This harmony lasts merely one measure before subsequently moving upward a half-step. Moreover, the ascription of a B major key center here ignores the larger-level bass ascent that controls the entire passage. In fact, the four subsequent downbeats following R. 9 contain a chromatic bass ascent from D# through G. The bass motion temporarily reverses direction, and is interrupted with competing material before returning to its ascent. At R. 10, the bass moves upward chromatically, beginning with F, and reaching Ab three measures before R. 11. Successive downbeats then ascend from Ab to Bb to C before returning to a clear and unambiguous Db tonal center at Rehearsal 11. The stepwise bass ascent between Rehearsals 9-11 thus organizes this entire section, and shows an underlying linear organization at work in the passage.

Thus the notion of a B major tonal center is negated by its brevity and by its relation to the larger-level stepwise bass ascent. Additionally, Spratt's third tonal center of G minor is likewise nested within this bass ascent, and is subordinate to it. Moreover, while a G minor triad is given to the lower voices twice (at four measures after R. 9, labeled 3, and at four measures after R. 10, labeled 4), it is, like the surrounding harmonies, harmonized bitonally. The G minor triad in the lower voices is juxtaposed against different triads each time, first with an E major triad, and second with an F major triad. The bitonal harmonizations, as well as the placement of the G minor triads within the larger bass ascent, call into question the notion of a G minor tonal center.

In connecting tonal relationships to the notion of "aesthetic balance and proportions," Spratt evidently attempts to extend to the domain of harmonic

Example 1.5a: Rehearsal 8-11 of *Mouvement Symphonique (Rugby)*

Example 1.5a (Continued)

Example 1.5a (Continued)

Example 1.5a (Continued)

Example 1.5a (Continued)

Example 1.5a (Continued)

Example 1.5b: Harmonic reduction of Rehearsal 8-11

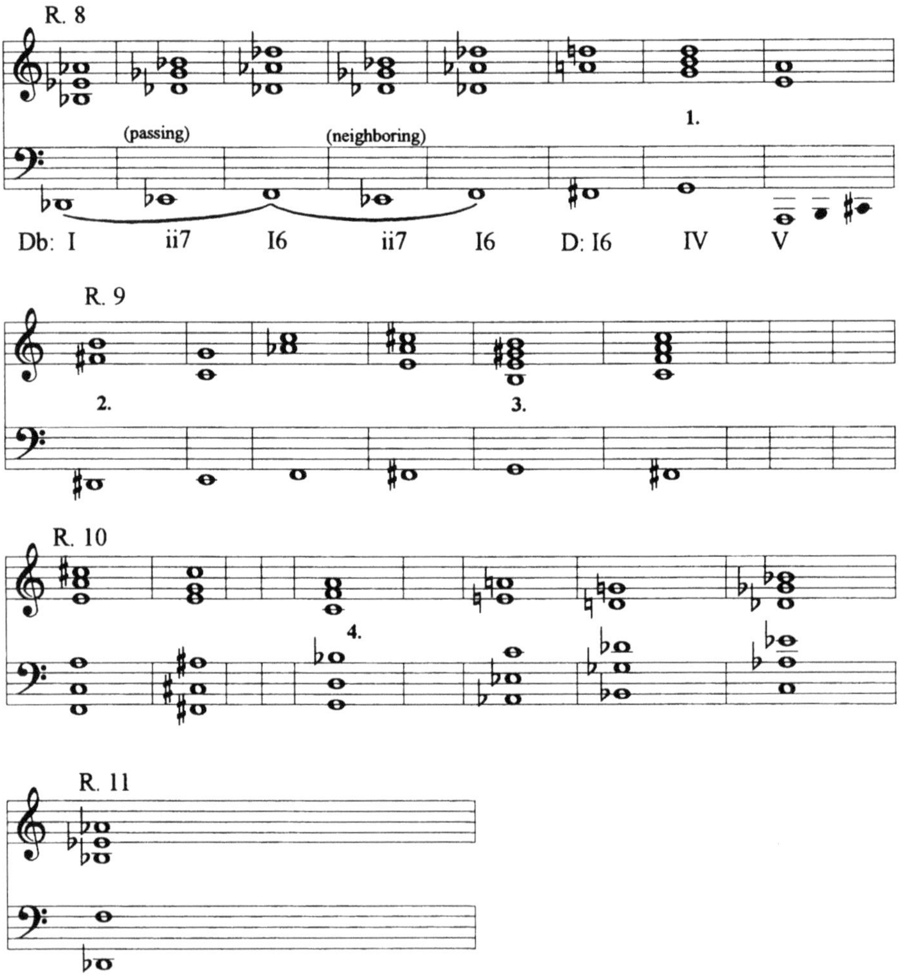

organization those ideas of formal and architectural symmetry addressed by earlier biographers and analysts. However, within the musical context, the identification of the tonal centers of G major, B major, and G minor seems to neglect both the larger contrapuntal implications of the passage and the harmonic function of the first eight measures.

Elsewhere in his study Spratt's observations about contrapuntal practice and strategy as a compositional factor in Honegger's output are important and compelling. In addition to pointing out the use of specific contrapuntal devices within the compositions, Spratt demonstrates that Honegger's concern for polyphonic, linear, and contrary-motion principles took primacy over harmonic organization early in Honegger's compositional career. Chapter 3 presents these aspects of contrapuntal design in greater detail.

Honegger as Author

Honegger's own writings reveal his stance on compositional matters. For thirty-five years, beginning in 1920, he frequently set forth in print his thoughts on compositional aesthetics, the role and profession of composer, and the difficulties and problems of publication and performance. He composed music for over forty films, and addressed in several places the relationship of music and film.[42] Much of his work as a music critic for the journal *Comoedia* was collected and reprinted in the book *Incantation aux fossiles*.[43] In addition, a series of conversations with critic Bernard Gavoty were printed under the title *Je suis compositeur*,[44] a book that discusses Honegger's background, influences, notational reforms advocated by the composer, reflections on the profession of composition, and—crucial for the present study—compositional methods and aesthetics. Throughout, his writings balance an ethos of craftsmanship with a willingness to respond to, or initiate, contemporary polemics.

Musical architecture was paramount to Honegger's compositional ethos, especially in his advocacy of formal symmetry by the use of reversed recapitulation and arch form, noted by numerous biographers and analysts. He addressed in detail his own ideas on compositional architecture in 1929, developing the architectural metaphor further by contrasting formal symmetry with the asymmetry of traditional thematic appearances in sonata form:

---

[42] See Honegger, *Ecrits*, especially the articles "Du cinéma sonore à la musique réele" (pp. 105-111); and "Particularités sonores du film 'Rapt'" (pp. 139-44).

[43] Arthur Honegger, *Incantation aux Fossiles* (Lausanne: Editions d'Ouchy, 1948). Honegger's music criticism written in the 1920s for the journal *Musique et Théâtre* have recently been collected and published in Honegger, *Ecrits*.

[44] Honegger, *Composer*.

> [T]he re-exposition of the order *A followed by B* is illogical, showing as it does, a defect in symmetry. In architecture you have a façade, the two left pillars of which, for instance, are arranged so that *A is followed by B*. Your object is solidly and harmoniously to rest the architrave and the pediment upon these pillars and upon those that form a pendant to them on the right. Symmetry as regards those on the right, will inevitably give you *B followed by A*, not vice versa, as the first time. Very well; in the sonata the pillars correspond to the themes; the pediment is the central development. Symmetrically, then, you must have *B followed by A* in the recapitulation since you had *A followed by B* in the exposition.[45]

Despite the importance of formal symmetry for the composer, numerous commentators disagree over the form of many of the works. The composer's writings on the subject suggest an *ad hoc* and contextual solution to form, especially sonata form. For Honegger, formal construction rejects prescriptive approaches and avoids the derivation of form from previous models or templates. A passage from *Je suis compositeur* reflects this idea:

> Honegger: The musician must first invent his model and reproduce it afterwards. If I desire to compose a sonata for piano and violin, I have absolutely nothing before my eyes or in my memory. I must invent everything.
>
> Gavoty: Do not other sonatas for piano and violin furnish you with models?
>
> Honegger: One cannot be satisfied with reproducing the form. Another sonata may, indeed, serve as a model, but the important thing is the actual sound, the themes, the melodies, the rhythms. If I copy, I am a conscious imitator, but in vain ... One must invent a personal model in the abstract, and construct it ideally. But this model will have no definite form before it is realized, for according to the material employed, the model will shift its form.[46]

Honegger's comments suggest that he does not base his appeals to musical tradition upon a rote cultivation of form. Instead, the dynamics of the musical materials suggest an ideal form. Other writings further suggest that thematic content strongly determines musical form. He develops an analogy with literature, stating:

> I frequently compare a symphony or a sonata with a novel in which the themes are the characters. After we have made their acquaintance, we follow their evolution, the unfolding of their psychology. Their

---

[45] Originally appeared in Paul Landormy, "Arthur Honegger" (translation by Fred Rothwell), *The Musical Times* (September 1929): 789. Translated by Spratt, *Honegger*, p. 266.

[46] Honegger, *Composer*, p. 68.

> individual features linger with us as if present. Some of these characters arouse feelings of sympathy, others repel us. They are set off against one another or they join hands; they make love, they marry or they fight.[47]

While privileging thematic content in his conception of sonata form, Honegger also dismisses a compositional idea typically found in prescriptive compositional approaches to sonata form: the notion of transitional or bridge themes that are subordinate to two primary themes. Instead, his polemic establishes an egalitarian thematic structure in which any number of themes may occur, each potentially equal in importance.

> Honegger: Let us take for example the bi-thematic movement of the sonata. The treatises teach that this movement shows two themes confronting one another. But what no one says is that between the first and second theme there may be a multitude of others. Thus, in analysing the String Quartet, Op. 59, No. 1, of Beethoven, I find nine absolutely distinct themes. The difficulty is resolved by saying: there are the themes, the transitional episodes, the—
>
> Gavoty: The bridges.
>
> Honegger: These are euphemisms. Why has not a given motif—under the pretext that it is the fifth or sixth—the right to the noble name of theme? Because the sonata is said to be bi-thematic? It is absurd.[48]

Despite the frequent adoption of sonata form in his instrumental works, and his self-styled posture as a traditionalist, Honegger repudiates many of the traditional concepts of sonata form. In dismissing a thematic structure linked by transitional themes, he contrasts to this the abstract values of "equilibrium" as well as a compositional ideal of what might be labeled "seamlessness" in the music:

> To me, a symphonic work must be built logically, without the possibility of interjecting the slightest anecdotal element between its different parts. I repeat: one must give the impression of a composition in which all is linked, the image of a predetermined structure. It has been said that architecture is frozen music. I should rather say that it is a geometry in time. Here, as elsewhere, one must be very exacting, so as to achieve an absolute equilibrium. If certain composers were the architects of their own houses, or builders of their own motor-cars, they

---

47 Ibid., p. 74.

48 Ibid., p. 69. Naturally, Honegger is responding here to contemporary views of sonata form as bi-thematic.

would long ago have been reduced to rubbish.[49]

Underlying Honegger's rejection of the notion of transitional sections and themes is an even more radical rejection of another aspect of traditional sonata form. Absent from his discussion is the consideration of form as expressing a *harmonic structure*. In fact, the notion of a predetermined tonal structure, traditional or otherwise, is refuted by the composer. In discussing the process of compositional development within a composition, we read:

> Gavoty: In this elaboration of your work, does the tonal design play a major role? For Vincent d'Indy, this plan played a determining role.
>
> Honegger: To me, the 'tonal design' is an abstraction. I do not even understand what it means. I have had numerous friendly debates with d'Indy himself on this topic. I told him at the time that to me the dogma of tonality seemed an entirely outmoded notion. As well require a contemporary dramatist to observe the unity of place of the classical tragedy, which so many masterpieces cheerfully ignore! ... Not everyone has 'absolute pitch.'
>
> Gavoty: It appears that Franck set down the tonal plan for his work at the top of his manuscript for his Symphony. It is said that he had decided upon it before having written a single note or jotted down the least theme.
>
> Honegger: It is a technique completely foreign to me, though I find it absolutely legitimate if the work benefits by it.[50]

These comments have several implications. First, Honegger is rejecting outright the notion of a *predetermined* tonal scheme for his own music. Secondly, he refutes broadly the traditional principles of tonality, both as supplying the melodic and harmonic language, and its role in enhancing, articulating, and structuring musical form and design. Thus, given the enhanced role which musical architecture seems to play for the composer, the less easily defined role of tonality in structuring musical form and design seems to create a vacuum, and we are left to ask: what organizing principles take its place?

Certainly Honegger was forthright in rejecting the uncritical adherence to compositional ideologies of various types, rejecting not only that of tonality, but of others as well:

> I cannot conceive of music fabricated by laws set up in advance. I am neither polytonalist, nor atonalist, nor a dodecaphonist. It is true, our contemporary musical material is based on a ladder of twelve chromatic sounds, but used with the same freedom as are the letters of the

---

[49] Ibid., 79.

[50] Ibid., 82-3. There are a few problems in reconciling this quotation with Honegger's compositional practice; those ideas are explored further in Chapter 6.

alphabet by the poet or the prismatic colors by the painter.[51]

Honegger's writings suggest that we look elsewhere than in a predetermined harmonic/pitch language for underlying organizational principles in the music. We find a key from his writings as early as 1920 in the way he fashioned an individual stance even while acknowledging his ties and indebtedness to musical tradition.

> I do not, like certain musicians, look for a return to harmonic simplicity. I find, on the contrary, that we ought to use the harmonic materials created by the school which has preceded us, but in a different sense and *based on lines and rhythms*.[52]

Although here Honegger acknowledges the influence of previous composers, at the same time he consciously sets himself apart from those influences through the idea of articulating musical form through the linear and the rhythmic dimension. These two elements were central to his musical thought throughout his entire career. In 1951 he was still advocating these two dimensions while rejecting tonality:

> ... it seems to me that we can no longer tolerate this fetishism of tonality, which has been a burden on entire generations of musicians. Of much greater importance than tonal equilibrium are *melodic and especially rhythmic balances*.[53]

Thus, despite Honegger's emphasis upon historical continuity and traditionalism, his writings nevertheless deny several traditional elements of musical composition, most notably tonality and traditional thematic function within sonata form. For the composer, the conventional functions of tonality are replaced by the emphasis upon formal symmetry and the notion of musical form as activated not by tonal structure, but what might be considered contrapuntal and rhythmic structures. Several earlier biographers and analysts have suggested that rhythm and counterpoint provide a foundation for understanding musical organization in the composer's work, but have not developed these ideas in significant detail. The following chapters will examine and establish how rhythmic and contrapuntal structures activate Honegger's compositions.

---

51 Ibid., 83.

52 Arthur Honegger, *Ecrits*, 34. Italics this author's.

53 Honegger, *Composer*, 83. Italics this author's.

Chapter 2

Rhythmic Structures

The Problem of Rhythm?

Honegger's writings surveyed in the previous chapter show an emphasis on rhythm as a critical aspect of compositional design, to the extent that this element takes precedence over tonal structure and organization. Rhythm figures prominently in the published historical and analytical studies that address primary stylistic features of Honegger's music. One general study describes Honegger's style as characterized by a "highly accented, but basically regular, even motoric rhythm."[1] In a metaphorical assessment, Pierre Meylan calls attention to the primacy of rhythm, which he claims is derived from the union of the rhythm of nature and of the machine.[2] Daniel Parker concurs, noting for the symphonies that "[r]hythm is an extremely important stylistic factor in these works. The consistent rhythmic energy that permeates these works is readily apparent to the listener, thereby establishing this factor's significance in the compositional hierarchy."[3]

Yet the language used to evoke this musical kineticism suggests a relentless mechanical or even machinelike regularity lacking in rhythmic nuance. Parker's statistical study perpetuates this notion, claiming that 96% of all the measures in the symphonies contain "unified, regularly-recurring pulse groupings." Of these, 94% display pulse groupings that conform to the meter, and 2% contain pulse groupings at variance with the prevailing meter signature. The residual 4% contain implied polymetric passages.[4] Similarly, Laurent Fichet discusses this aspect of metric regularity in the symphonies. Unlike other analysts, Fichet also extends his observations to higher levels of rhythmic structure, implicitly engaging the notion of hypermeter when he asserts that the first movement of the Fifth Symphony groups primarily into four- or eight-measure units.[5]

---

1 Eric Salzman, *Twentieth-Century Music: An Introduction* (Englewood Cliffs, N. J.: Prentice-Hall, Inc., 1974), 59.

2 Pierre Meylan, *Arthur Honegger: Humanitäre Botschaft der Musik* (Frauenfeld and Stuttgart: Verlag Huber, 1970), 204.

3 Daniel Quienton Parker III, "The Symphonies of Arthur Honegger: An Examination of Stylistic Factors and Form" (Ph.D. diss, University of Kentucky, 1990), 233.

4 Ibid., 191.

5 Laurent Fichet, "Problèmes de langage dans l'oeuvre symphonique d'Arthur Honegger," in *Honegger-Milhaud: Musique et Esthétique, Actes du Colloque International tenu à*

Is there a "problem" of rhythm in Honegger's music? The tyranny of the barline implied by Parker and Fichet seems to suggest one, as do the descriptions of the music as motoric or machinelike. Possibly this is a result of the limited repertoire that these writers address. Additionally, passages in the symphonies that do upset the metric regularity may take on a heightened formal or structural role.[6]

We may find a reconciliation by considering global organizational aims, a focus lacking from Parker's and Fichet's studies of the symphonies. Kurt von Fischer provides a partial solution to this dilemma posed by the rhythm problem: "Rhythm is to be understood not only as a vitalizing component, but also as a formal one, inasmuch as similar or various contrasted rhythmic patterns determine the character of whole sections and even entire movements."[7]

In this sense, then, rhythmic organization provides a means of formal delineation and contrast, and suggests organizational strategies that may structure sections, movements, and compositions. At this larger level, rhythmic organization, far from supplying an unyielding mechanical pulse, provides a powerful source for musical conflict and resolution. For Honegger, this dimension plays a crucial role in the construction of large musical forms, usurping the role played by tonal organization in music of the eighteenth and nineteenth centuries.

Rhythmic organization thus takes on a greatly expanded role in the compositions, both in large-scale and in smaller-scale organization.[8] The following analyses will show how these strategies operate in overall compositional design, showing processes of rhythmic crescendo and metric organization in *Mouvement symphonique* (*Pacific 231*), *Sonate pour violon*, and *Prélude pour "La Tempête"*. Further consideration of the latter two works point out how the composer articulates rhythmic relationships across different layers of rhythmic structure, such as that of rhythmic subdivision and meter. Moreover, the composer develops the same rhythmic conflicts across *different layers of rhythmic structure* within a composition, generating multi-leveled rhythmic correspondences and relationships. In observing techniques of rhythmic conflict and the relationships engendered across rhythmic and metric layers, the

---

la Sorbonne en novembre 1992 réunis et édités par Manfred Kelkel (Paris: Librairie J. Vrin, 1994), 118-19.

6 Elsewhere, however, Parker's data imply a dimension of rhythmic complexity through the use of syncopation, which he claims appears in 35% of all the measures of the symphonies. See Parker, "The Symphonies," 198.

7 Kurt von Fischer, "Arthur Honegger," chap. in *Essays in Musicology*, ed. Tamara Evans and trans. Tamara S. Evans (New York: The Graduate School and University Center, City University of New York, 1989), 164.

8 Honegger extends purely rhythmic organization even to the vocal works, some of which employ rhythmically-notated speech, such as the opera *Antigone*, as well as the oratorios *Cris du monde* and *Danse des morts*.

discussion is motivated in part by terminology and issues germane to the issues of rhythmic consonance and dissonance, and proposes certain refinements and elaborations of these ideas.

Preliminaries: Compositional Sketches of *Mouvement symphonique* (*Pacific 231*)

Rhythm takes on its most enhanced formal and structural role in Honegger's three *Mouvements symphoniques*. Honegger's compositional sketches illuminate some of his rhythmic techniques and compositional processes. The following examples from his sketches show Honegger's work with individual motives, and two particular techniques come to light: first, the use of rhythmic crescendo; and second, the construction of rhythmic content *prior* to pitch content. After pointing out the application of these two approaches to individual motives, the analyses will then show their use in an entire composition.

Rhythmic Crescendo. Rhythmic crescendo refers to a developmental process that makes use of increasingly smaller rhythmic values. A motivic fragment contained in the compositional sketches for *Mouvement symphonique* (*Pacific 231*) makes its appearance not in the first, but in the second *Mouvement symphonique* (*Rugby*), published five years later in 1928. The motive in Example 2.1, from the printed score, reveals the process of acceleration characteristic of rhythmic crescendo. In the example, there is a shift to a quicker rhythmic value with each change in instrumentation: the motive is stated in eighth notes in the cellos, transposed and recast into eighth-note sextuplets in the violas, then sixteenth notes in the first violins. Even as the rhythmic values quicken, Honegger preserves and emphasizes the identity of the motive by accenting the beginning of each six-note grouping.

The rhythmic transformations on the six-note motive have implications for the perception of the meter. The initial eighth-note organization of the six-note motives (played by cellos, mm. 57-60) conforms to the barline. However, the preservation of the six-note motives upon the move to eighth-note sextuplets (played by violas, mm. 61-64) and sixteenth notes (played by violins, mm. 65-66) now challenges the 3/4 meter. Here it is possible to for the listener to hear the notated barline give way to a perceived barline, corresponding to the accented beginnings of the six-note motives. In this case, not only do the underlying rhythmic values accelerate, but the pacing of the perceived barline may accelerate accordingly, first stratifying a layer of 6 eighth-note sextuplets, then stratifying a layer of 6 sixteenth notes.[9]

---

[9] Many of the subsequent analyses will point out motivic groupings that are in conflict with the notated meter. The repetition of motives of a consistent length may potentially bring about the perception of a new meter that is determined by the length of

Example 2.1: M. 57 ff.; *Mouvement symphonique, #2 (Rugby)*

Despite these potential threats to the notated meter, it is characteristic of Honegger to keep the original notated meter intact throughout. In *Je suis compositeur*, Honegger matter-of-factly dismissed the practice of altering the

---

these motives. Whether or not listeners perceive a new meter when musical cues suggest this type of conflict is dependent on any number of factors, including musical context, the brevity or length of the passage in conflict with the underlying meter, how strongly or weakly the notated meter is established prior to the conflict, performance decisions, a listener's familiarity with the composition, or a listener's overall experience with music. Andrew Imbrie has suggested that listeners hypothetically fall into two categories. On the one hand, "conservative" listeners will maintain as long as possible a previously established meter in the face of conflicting metric cues; on the other hand, "radical" listeners will quickly and immediately shift their metric perception to the new implied meter. See Imbrie's essay " 'Extra' Measures and Metrical Ambiguity in Beethoven," contained in *Beethoven Studies*, ed. Alan Tyson (New York: W.W. Norton and Co., 1973), 45-66. Given this, Wallace Berry's rebarring of particular musical passages presumes a radical listening strategy: see Wallace Berry, *Structural Functions of Music* (Englewood Cliffs, New Jersey: Prentice-Hall, 1976). Joel Lester's *The Rhythms of Tonal Music* (Carbondale: Southern Illinois University Press, 1986) presumes a more conservative listening approach. Lester itemizes the musical conditions that create accent (harmonic change, textural accent, contour accent, pattern beginnings, etc.) and shows how certain passages of music achieve their rhythmic vitality by the rich interaction of accent and underlying metric structure.

These types of conflicts have been explored at length through the notion of rhythmic (or metrical) consonance and dissonance. See Maury Yeston, *The Stratification of Musical Rhythm* (New Haven and London: Yale University Press, 1976), and Harald Krebs, "Some Extensions of the Concept of Metrical Consonance and Dissonance," *Journal of Music Theory* 31/1 (1987): 99-120, and Krebs, *Fantasy Pieces: Metrical Dissonance in the Music of Robert Schumann* (New York and Oxford: Oxford University Press, 1999). Krebs also provides the useful distinction between "direct" metrical dissonance, in which two conflicting metrical layers are stated simultaneously, and "indirect" metrical dissonance, in which two conflicting layers are stated successively.

The analyses here do not mean to suggest that these metric conflicts *necessarily* create the perception of a new meter. Instead, motivic groupings at odds with the meter create an expressed rhythmic or metric layer that may or may not result in metric reorientation on the part of the listener.

musical meter and barring passages in a manner consistent with motivic groupings rather than with an underlying meter:

> Finally, I should like to call the attention of budding composers to a custom which is already terribly dated: that of a perpetual and excessive changing of time. A number of great masters have made use of this and abused it. They confuse rhythm with prosody. Time should play the role of the highway mile-post. To lengthen or curtail these, according to the changes in landscape, would be nonsense. Why deliberately complicate the performance? ... The most complex rhythms can be indicated within the frame of two, three, or four beats per measure. The performance can only be improved, without forcing these burdensome calculations upon the interpreters, for they destroy all feeling of freedom. Furthermore, the kind of writing in which 7/32 time is blended into 5/1 time does not dazzle anyone. Therefore it is unnecessarily pretentious.[10]

Rhythmic Motives Constructed Prior to Pitch Content. Sketch material for Honegger's first *Mouvement symphonique* (*Pacific 231*) suggests that the composer sometimes fixed the rhythmic organization of particular motives prior to establishing the pitch organization. In these instances at least, rhythm precedes pitch in the compositional process. Example 2.2a contains a preliminary version of the motive for Rehearsal 7 from the sketches, while 2.2b shows its appearance in the composition.

A comparison of the two indicates that Honegger retains the overall rhythm and the contour from the sketch to the published version. The motive in Example 2.2b shifts in the barline, moving ahead by one beat. In addition, it is extended by two eighth notes. However, note that the pitch content of the preliminary sketch motive at 2.2a is significantly different, and is more tonally oriented than the later published version. In the sketch version at Example 2.2a the first four pitches state a major triad (E-G-C-E) while, in contrast, those of the published composition assert an [0156] tetrachord (F-B-E-C). The example shows Honegger fashioning the overall rhythmic design of the motive first, supplying pitch material that was workable but that required later revision.

Example 2.2a and 2.2b: Motive from sketch and composition

2a. Motive from sketch   2b. Motive from composition

[10] Arthur Honegger, *I Am a Composer*, trans. by Wilson O. Clough in collaboration with Allan Arthur Willman (New York: St. Martin's Press, 1966), 57-8.

Rhythmic Organization: *Mouvement symphonique* (*Pacific 231*)

Honegger's compositional sketches further reveal that he used both these techniques—rhythmic crescendo and the development of rhythmic content prior to pitch content—not only in relation to individual motives, but also to an *entire* composition, his *Mouvement symphonique* (*Pacific 231*). The work is Honegger's most evident example of a composition determined by rhythmic structure. Critical reaction to this work focused upon its onomatopoeic aspect, since its overall process of rhythmic crescendo gave the musical effect of a steam locomotive, the *Pacific 231* of the subtitle. However, Honegger claimed that the subtitle was an afterthought, and he stressed the abstract and absolute rather than the programmatic nature of the composition:

> To tell the truth, in *Pacific* I was on the trail of a very abstract and quite ideal concept, by giving the impression of a mathematical acceleration of rhythm, while the movement itself slowed. Musically, I composed a sort of big, diversified chorale, strewn with counterpoint in the manner of J. S. Bach.[11]

A precompositional plan appears within the compositional sketches for the work. This plan is based solely upon rhythmic organization, showing at a preliminary stage the process of rhythmic crescendo that was to structure the composition. The absence of any pitch material in the plan suggests that the rhythmic dimension of the composition was most likely determined prior to the pitch content. Example 2.3 displays the plan: on the left, the rhythmic unit is given; on the right, the amount of measures for that rhythmic unit to occupy.[12] Following the initial whole note unit, the successive subdivisions can be understood to partition three overlapping layers of rhythmic structure: division of the measure (half note, half-note triplet, quarter note), half-measure (quarter note, quarter-note triplet, eighth note), then at the beat (eighth note, eighth-note triplet, sixteenth note). Thus on each layer occurs an arithmetic increase of subdivisions where measure, half-measure, and beat are successively divided by 2, 3, and 4. In the sketch, the penultimate row contains a mixed level of subdivisions, and the final row a return to sixteenth notes.

---

[11] Honegger, *Composer*, 101. Halbreich notes some musical correspondences between this work and Honegger's earlier work for the music for Abel Gance's film *La Roue*, which used the locomotive as a subject (Halbreich, *Honegger*, trans. Nichols, p. 351). In light of this, Honegger's denial of any programmatic content for *Pacific 231* suggests a decided ambivalence toward acknowledging any programmatic origins to this work. A similar issue exists with Honegger's *Symphonie pour cordes*. This is discussed briefly here in Chapter five.

[12] Reconstructed from sketches at the Paul Sacher Foundation (Microfilm 186: 785). Numbers were not supplied to the right-hand column for the fourth row (quarter-note division) nor for the penultimate row.

Example 2.3: Precompositional sketch for *Mouvement symphonique*

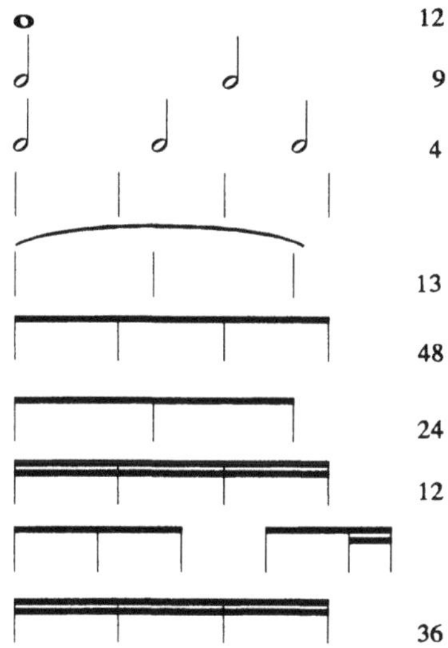

In the printed score, Honegger retained the overall process of rhythmic crescendo, but expanded the dimensions of his precompositional sketch. Figure 2.1 shows the sections for the printed score determined by the primary rhythmic value, as well as the metronome markings that correspond to the rhythmic shifts. The metronome markings following no. 7 (slowing from quarter note = m.m. 152 to 132) show the overall deceleration taking place beneath the increase of rhythmic values—this is the process that Honegger describes in his quote above.

A comparison of the precompositional sketch (Example 2.3) with the structure of the printed score in the following Figure 2.1 reveals some differences. With few exceptions, the measure lengths of the sections differ. Only the half-note triplet (no. 3 of Figure 2.1) retains its length of four measures from the precompositional sketch.

Additionally, in the printed score Honegger has added the level of dotted half note (this is no. 2 of Figure 2.1). With this change, Honegger now extends the arithmetic division by 2, 3, and 4 (of measure, half-measure, and beat) of the precompositional sketch to a higher level of organization, the two-measure level. Thus, nos. 1-3 of Figure 2.1 bisect two-measure groupings: the whole note divides the two-measure unit in two, the dotted half divides this unit in three, and the half note divides the two-measure grouping in four. Note that this additional level of dotted half note (no. 2) has a potential impact on the metric structure of mm. 14-16: the dotted-half note groupings may be heard as implying a meter of 3/4 in these measures.

Figure 2.1: Rhythmic structure of *Mouvement symphonique*

Introduction	m. 1	(Half note = 60)
1. Whole note	m. 12	(Half note = 80)
2. Dotted half	m. 14	
3. Half	m. 17	
4. Half-note triplet	m. 23	
5. Quarter note	m. 27	
6. Quarter-note triplet	m. 62	
7. Eighth note	m. 73	(Quarter note = 152)
8. Eighth-note triplet	m. 118	(Quarter note = 144)
(M. 147: Quarter note = 138)		
9. Sixteenth note	m. 159	
(Chorale melody)	m. 169	(Quarter note = 132)
Coda	m. 204	

Rhythmic Crescendo in *Mouvement symphonique* (*Pacific 231*): Further Refinements

With both the precompositional sketch and the finished composition, rhythmic crescendo provides a systematic and overarching scheme for Honegger's composition. In the finished composition, the acceleration process moves from whole note to sixteenth note, creating divisions of the measure that extend from 1 to 16. In addition, *Pacific 231* makes a further refinement to this process as the composition moves from quarter note (m. 27), to quarter-note triplet (m. 62), through eighth note (m. 73).

These three rhythmic values divide the measure by 4 (quarter note), 6 (quarter-note triplet), and 8 (eighth note). The composition further makes use of *intermediary* divisions, first between the divisions of quarter note and quarter-note triplet (between measure divisions of 4 and 6), and between the divisions of quarter-note triplet and eighth note (between measure divisions of 6 and 8). Honegger accomplishes this in several ways.

Example 2.4 contains mm. 54-76 of *Pacific 231*, beginning with the final eight measures of the quarter note section of the composition. Already here quarter-note triplets (played by the horns) appear in mm. 54, 55, 57, 58, 59, and 61, anticipating the following section, and thus blurring the sectional division at m. 62. In addition, though, the quarter-note quintuplets in mm. 56 and 60 play an important role: they provide measure divisions of 5 between the measure divisions of 4 supplied by the quarter note and measure divisions of 6 supplied by the quarter-note triplet. Thus, by providing an intermediate rhythmic value between quarter note and quarter-note triplet, these quarter-note quintuplets participate in the systematic acceleration of rhythmic values.

Example 2.4: *Mouvement symphonique*

Yet Honegger does not abandon groupings of 5 at the sectional change at m. 62. Instead, he states them on a *different layer* of rhythmic structure. Although mm. 62-66 are subdivided by quarter-note triplets (however, note that the upper score at m. 62 shifts to 6/4 meter and reinterprets the previous half note as dotted half note: quarter-note triplets are now indicated as quarter notes from mm. 62-66), the expressed groupings in vlns. 1 and 2 organize the motives into groupings of 5, articulated by motivic patterning, instrumentation, and accent. Compositional sketches for this passage, in fact, bracket the five-note motives, labeling them A and B, alternatively. Thus Honegger creates an organizational layer of 5 which reflects back to the quarter note quintuplets of mm. 56 and 60, but which become stated in a distinctly different manner. At mm. 56 and 60, the barline remained inviolate despite the quarter-note quintuplets. In contrast, the motivic groupings in the violins between mm. 62-66 create an organizational layer of 5 that *does not correspond but conflicts* with the barline. The composer therefore creates organizational layers of 5 in two ways, first preserving the barline with the quarter-note quintuplets (mm. 56 and 60), then disrupting the barline organization with the motive groupings of five notated quarter notes (mm. 62-66).

This conflict between the notated meter and the motivic groupings of 5 is expressed explicitly at mm. 62-66. The basses (bass clef), descending chromatically in half notes, uphold the barline organization, and may be heard to group the notated quarter notes into a layer of 6. On the other hand, the violins (treble clef) assert a layer of 5 notated quarter notes in conflict with the notated barline. The passage continues until these two conflicting layers of 5 and 6 share a notated metric downbeat at m. 67: this is determined by the least common multiple of the violins' layer of 5 and the basses' layer of 6.[13]

Overall, from mm. 54-66, the process of rhythmic crescendo is far richer than merely a rote mechanical shift from measure subdivision by 4 (quarter note) to subdivision by 6 (quarter-note triplet). Not only does Honegger anticipate the upcoming rhythmic value of quarter-note triplet between mm. 54-61, he creates an intermediary level of subdivision by 5, stating quarter-note quintuplets at mm. 56 and 60. Furthermore, between mm. 62-66 he establishes groupings of 5 in the violins that conflict with the barline. Additionally, between the upper and the lower parts he creates *simultaneous* layers of 5 and 6 by the conflicting groupings of the notated quarter note.

We may hear a similar process underway in the measures that follow. While the acceleration to eighth notes occurs at m. 73, the preceding passage from mm. 67-72 may be heard as transitional, linking quarter-note triplet

---

[13] Gretchen Horlacher refers to this least common multiple brought about by two conflicting motivic groupings as a "cycle;" see Gretchen Horlacher, "The Rhythms of Reiteration: Formal Development in Stravinsky's Ostinati," *Music Theory Spectrum* 14 (1992): 171-87.

organization (measure subdivision by 6) with the upcoming eighth notes (measure subdivision by 8). Note that the trumpet states eighth notes in mm. 67, 69, and 70, anticipating the new subdivision. In addition, the shift from measure subdivision by 6 to 8 is mediated by the groupings of 7 stated by the trumpet at mm. 68, 71, and 72.

Even after the move to eighth-note organization takes place at m. 73, Honegger still maintains a layer of 7, created by the motivic organization in the violins and emphasized by the accentual structure. This layer of 7 recalls the eighth-note septuplets of mm. 68, 71, and 72, but—like the groupings of 5 in the previous discussion—this layer now becomes stated on a different layer of rhythmic structure. The mm. 68, 71, and 72 septuplets preserved the barline organization, but the motivic groupings of 7 between mm. 73-75 conflict with the notated barline.

As in mm. 62-66, this conflict between mm. 73-75 is stated explicitly. Again the chromatic descent in the basses preserves the barline organization and thus groups the eighth notes into a layer of 8. In contrast, the upper part asserts a conflicting organization, and provides a layer of 7 eighth notes that challenges the barline. (Note that here, unlike the passage between mm. 62-66, Honegger does not maintain this section until the two conflicting layers of 7 and 8 share a notated downbeat. Instead, the octave descent in the bass, notated as a descent from Ab-G#, seems to dictate the length of the passage. The final statement of the violin motive at mm. 75-76 is shortened.)

In sum, the section between mm. 67-76 shares many of the same strategies as mm. 54-61. The move to eighth note organization at m. 73 is anticipated in mm. 67, 69, and 70. The eighth-note septuplets at mm. 68, 71, and 72 mediate with 7 the shift from measure subdivision by 6 (quarter-note triplet) to subdivision by 8 (quarter note). After stating the groupings of 7 that conform to the barline, Honegger then creates motivic groupings of 7 between mm. 73-75 that conflict with the barline. Finally, note that the upper and lower parts between mm. 73-75 simultaneously organize the eighth notes into conflicting layers of 7 and 8.

The overall process of rhythmic crescendo that structures *Pacific 231* is patently clear, and his precompositional sketch suggests that Honegger worked out this process in advance of his composition. However, other more subtle rhythmic relationships are at work in the completed composition, especially evident at mm. 54-76, which moves from the rhythmic subdivision of quarter note through quarter-note triplet to eighth note. The anticipation of upcoming rhythmic values masks sectional divisions at mm. 62 and 73. Further, through the use of quarter-note quintuplets and eighth-note septuplets, Honegger provides intermediary measure subdivisions of 5 and 7 between those of 4, 6, and 8. Finally, the composer creates rhythmic relationships across different layers of rhythmic structure: the groupings of 5 and 7 are stated in two different ways, either conforming to the barline or conflicting with the barline. All these strategies show a highly nuanced approach to rhythmic crescendo.[14]

Metric Organization

Rhythmic crescendo provides one technique for organizing an entire composition. Honegger makes use of other overall strategies for creating compositional structure in the rhythmic and metric dimensions. Following a brief overview of analytical approaches to overall metric organization, the following analyses will show how large-scale metric organization operates as a constructive device in two of Honegger's compositions.

Theorists of tonal rhythm have advanced the thesis that metric strategies provide global compositional organization analogous to the background tonic triad in the domain of pitch. Harald Krebs pursues this pitch/rhythm analogy by stating that, once established, the "primary metrical consonance"—usually usually represented by the meter signature—is conceptually maintained by the listener or performer even when not apparent, or even when contradicted. Richard Cohn draws this Schenkerian analogy even tighter in claiming that "the primary metric consonance is subject to *prolongation* through *displacement* by a meter with which it is dissonant." In another analysis, Cohn examines the recursive nature of meter and hypermeter, suggesting that the successive moves between duple and triple metric and hypermetric states provide the dramatic content of the Scherzo from Beethoven's Ninth Symphony.[15]

For these analysts, departure-and-return models of metric organization indicate global strategies that organize a movement of a composition. In the instance of Honegger's early unpublished *Sonate pour violon et piano en ré mineur* (1912),[16] we may observe a similar process not only at the level of a single movement, but operating across movements of the composition as well. Within the first movement of the sonata the duple 4/4 meter of the first theme section is supplanted by the triple 9/8 of the second theme. This creates formal delineation through both metric and thematic means. In addition, as the cyclic

---

[14] Halbreich refers to these passages as the "unstable" passages of the composition, and notes that this is "an early, successful example of the principle of metric modulation found in the work of Elliott Carter and foreseen by Charles Ives." See Harry Halbreich, *Arthur Honegger*, trans. Roger Nichols (Portland, Oregon: Amadeus Press, 1999), 351.

[15] Krebs, "Some Extensions," 105; Richard Cohn, "Metric and Hypermetric Dissonance in the *Menuetto* of Mozart's Symphony in G Minor, K. 550," *Intégral* 6 (1992): 3; Richard Cohn, "Dramatization of Hypermetric conflicts in the Scherzo of Beethoven's Ninth Symphony," *19th-Century Music* 15/3 (1992): 22-40.

[16] Paul Sacher Stiftung (Microfilm 186: 566-612). Although this work is unpublished, it has been recorded on Vol. 1 of Arthur Honegger, *La musique de chambre*, CD 1C1008, Timpani, 1992.

organization thematically frames the entire work, so too the metric organization provides an overall strategy for the beginning and ending of the composition. The first movement begins in duple 4/4 meter and finishes in triple 9/8, while the Allegro of the final movement begins in triple, "resolving" to 4/4 meter at the close.[17]

Sonate pour violon et piano en ré mineur (D = Duple; T = Triple)

Movement 1	Movement 3
4/4-----9/8	3/4------4/4
(D------T)	(T--------D)

At furthest remove, then, meter and metric organization provide an overarching constructive element for the composition. Similarly, shifts in duple and triple metric states supply the formal organization for another composition, Honegger's *Prélude pour "La Tempête"*, an orchestral work published in 1923. Here the alternation of these metric states establishes the sections of the composition, which are defined by ostinati that assert either 2/2 or 6/4. The alternation of these two meters creates a duple/triple conflict at the level of the half-measure. Moreover, the motivic material stated against these ostinati often appears in the opposing meter, allowing for metric conflict not only through the *succession* but also through the *superimposition* of both the 2/2 and 6/4 meters.

Although motivic material recurs across the different formal sections, this material is alternately cast in the meters of 2/2 and 6/4. Significantly, the work maintains continuity in the motivic and thematic dimensions, but creates contrast and conflict through metric organization.

Example 2.5 contains the rhythmic figures that dominate the ostinati of the first four sections of the *Prélude pour "La Tempête"*. Sections A and D are defined by ostinati in 6/4, section C by 2/2. Section B thus acts as a metric transition between A and C. Both meters appear during the course of a single measure, but alternate within the course of that measure. 6/4 is expressed for the first half of the measure, 4/4 for the second half. The conflicting rhythmic identities of sections A and C are mediated by section B.

Between sections B and C, the composer reverses the metric terms of both the motivic and ostinato material. Example 2.6 shows that the 2/2 motivic material (upper staff) that begins section B recurs, transposed up a major seventh and rhythmically recast into 6/4, at section C. At the same time, conversely, the 6/4 ostinato of section B (lower staff) is replaced by the 2/2 ostinato in section C.

---

17 Following Cohn, I have collapsed metric organization into two states, duple and triple. Therefore pulse organization based upon exponents of 2 are considered duple, upon exponents of 3 are triple.

Example 2.5: *Prélude pour "La Tempête"*

Example 2.6: Sections B and C (mm. 23-25, 52-53)

A similar procedure occurs within section E, which forms the development section of the composition. Ostinato and motivic material are set and recast into opposing metric identities. This now takes place within the same formal section. Example 2.7a shows that the 2/2 ostinato figure that begins section E recurs nine measures later, now shifted to 6/4. Furthermore, Example 2.7b shows that the motivic material asserted in 6/4 at R. 7 returns in altered form seven measures later, now stated in 2/2.

Example 2.7a: Ostinato at Section E (mm. 69-80)

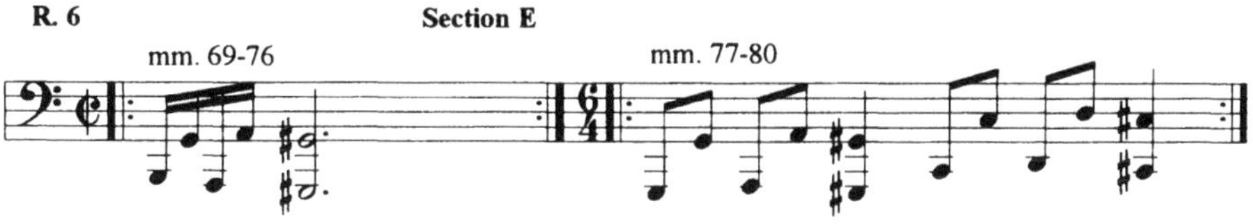

Example 2.7b: Motivic material at Section E (mm. 89, 96-97)

Thus, in the *Prélude pour "La Tempête"*, the conflicting duple and triple organization (stated at the level of the half-measure) both articulates formal divisions and provides a means for musical contrast and development within sections. Motivic and ostinato statements are simultaneously stated in different meters, then each is subsequently given the opposing metric identity and juxtaposed again.

In both the *Prélude pour "La Tempête"* and the *Sonate pour violon*, compositional organization is effected through large-scale metric strategies. In the *Sonate pour violon*, the move from duple to triple meter at the beginning of the composition's first movement is mirrored at the end of the final movement, which proceeds from triple to duple. In the *Prélude pour "La Tempête"*, metric shifts delineate form and participate in developmental processes. Despite the differences in formal structure and harmonic language of the two compositions, both works make use of shifts between duple and triple metric states, underlying the arch form and atonality of the *Prélude*, as well as the traditional sonata form and late-Romantic harmonic language of the early *Sonate*.

Other Correspondences in *Prélude pour "La Tempête"*: Polymetric Organization

Much of the *Prélude pour "La Tempête"* states two different meters simultaneously, articulating a conflict between one layer of 6 quarter notes and another layer of 4 quarter notes. It is useful to define two different strategies for polymetric organization in order to show how Honegger works out similar conflicts in different ways. Following Yeston and Krebs, these conflicts will be referred to as "dissonances."[18] One type involves conflict engendered through *conflicting subdivisions of the notated measure*. An abstract realization of this type of conflict is given in Example 2.8a. We can refer to this example of 3 : 4 dissonance as "measure-preserving," since the integrity of the barline is maintained.

Example 2.8b likewise manifests a 3 : 4 conflict, yet, in contrast, this "tactus-preserving" dissonance is brought about by *conflicting groupings of the notated beat*. Here the tactus is kept intact while the integrity of the barline is disrupted by the expressed layer of 3/4 in the upper stratum of Example 2.8b. Although these two types of dissonance are related abstractly—the accentual patterns of Example 2.8b compose out in augmentation those of 8a—they are distinguished by their orientation in relation to the tactus and to the barline, composing out the rhythmic conflict on different layers of rhythmic structure.[19]

The distinction between these two types of dissonance helps to elucidate polymetric correspondences and relationships within the compositions of Honegger. For example, the primary polymetric conflicts worked out in the sections of the *Prélude pour "La Tempête"* articulate the measure-preserving form of dissonance. The 6/4 and the 2/2 meters occur within the same time-span,

---

[18] See Yeston, *The Stratification of Musical Rhythm*; Krebs, "Some Extensions;" and Krebs, *Fantasy Pieces*.

[19] I would like to thank Steve Larson for helping to fashion the terminology for these two types of dissonance.

Example 2.8a: "Measure-preserving" dissonance

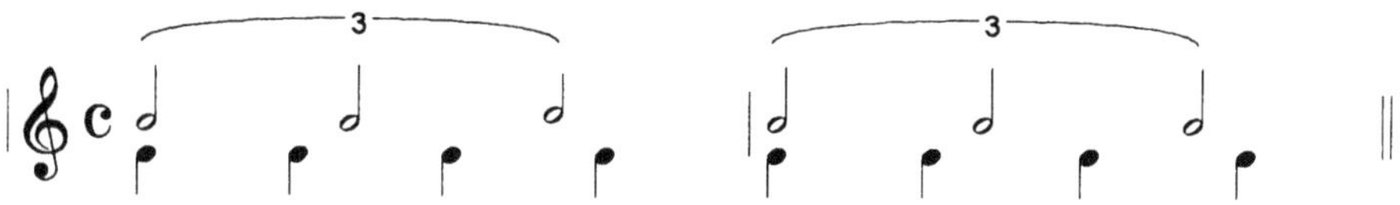

Example 2.8b: "Tactus-preserving" dissonance

each subdividing the measure differently. (See Exx. 2.5-2.7). Yet this is not the only mode of conflict explored in the composition. At the opening of the composition the melodic material suggests an expressed layer of four quarter notes stated against the written 6/4 of the ostinato. As in the body of the composition, the composer stratifies 4/4[20] against 6/4; however, the dissonance at the opening of the composition is tactus- instead of measure-preserving. Example 2.9 shows an abstract realization of the two types of dissonance for the composition.

Example 2.9: Tactus- and measure-preserving dissonance in *Prélude pour "La Tempête"*

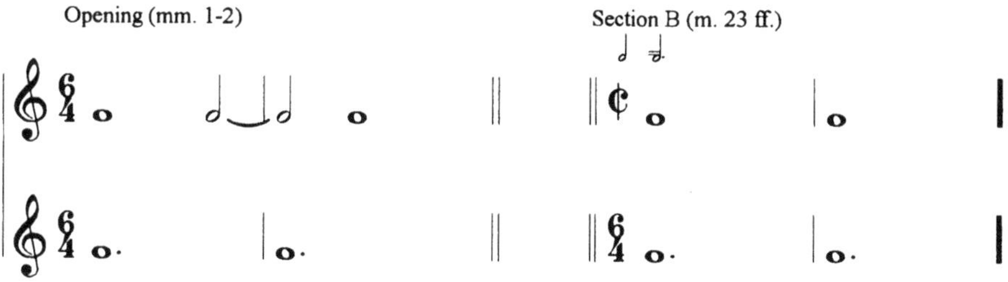

[20] However, the composer notates the duple meter in the body of the composition as 2/2 (cut time).

The example shows that the composer opens the composition expressing a 6: 4 polymetric conflict. In this instance, this is a tactus-preserving dissonance. This 6 : 4 conflict is maintained, but then becomes transformed within the body of the composition, and stated on a different layer of rhythmic structure. This provides an abstract rhythmic correspondence with the opening, and this polymetric conflict then supplies the underlying metric organization for the body of the composition.

Correspondences of Rhythmic Subdivision and Meter

The tactus- and measure-preserving dissonances contain polymetric relationships that are distinguished by their orientation to the notated beat and measure. In addition, Honegger develops correspondences across other layers of rhythmic structure. Examination of the thematic content of the first movement of the unpublished *Sonate pour violon et piano en ré mineur* (1912) suggests correspondences between the layers of tactus subdivision (divisions of the pulse) and meter.

Example 2.10 provides the beginning to the introductory theme (i), the first theme (Th1), and the second theme (Th2). The alignment of the three themes in the example shows how some rhythmic features of i carry over into Th1 and Th2. The dotted eighth and sixteenth notes of Th1 derive their metric position on the second beat from i. Similarly the eighth-note triplet of Th2 derives its metric position on the first beat from i. Consistently, the first theme area is defined by duple subdivisions, the second theme area by triple.

Example 2.10: *Sonate pour violon et piano en ré mineur*

These rhythmic subdivisions, moreover, project onto the *metric* layer. The introductory theme contains both duple and triple subdivisions. This duple/triple conflict stated within the introductory theme becomes composed out at the formal level. Between Th1 and Th2 the shift from duple to triple subdivision is magnified by the metric shift from the duple 4/4 to the triple 9/8 meter: the succession from duple to triple expresses the opening conflict at both the subtactus and the metric layer.

Correspondences between the subtactus and the metric layer contribute to the notion of rhythmic consonance, here generalized to a "rhythmic/metric consonance." In order to reveal these correspondences across rhythmic subdivision and meter, the "rhythmic/metric pair" groups tactus subdivisions on the left, and metric grouping on the right. Given this, the rhythmic/metric pairs for the opening sections of the sonata are:

Introductory Theme	(3,4)(4)
Theme 1	(2,4)(4)
Theme 2	(3)(3)

The relationship of subtactus layer to metric layer may be described as rhythmic/metric consonance in the case of a pure duple or triple identity.[21] This may be defined formally: for any rhythmic/metric pair (a)(b), rhythmic/metric consonance exists when, for all a and b, $a^x = b$ or $b^x = a$, and x is an integer. Rhythmic/metric consonance therefore requires the same identity—here either duple or triple—both at the level of tactus subdivision and at the level of metric organization. Rhythmic/metric consonance quarantines subdivisional and metric layers that are in an exponential relationship away from those that are not, preserving an uncontaminated duple or triple environment.

Thus the rhythmic/metric pairs above help show a fundamental organizational strategy that takes place at these three points in the sonata. The presence of both duple and triple subdivisions in the introductory theme prevents rhythmic/metric consonance from taking place. Rhythmic/metric consonance does occur at Theme 1 and Theme 2, however. Thus, the dissonance created through the juxtaposition of duple and triple subdivisions in the introductory theme becomes stated at the formal level, whereby Theme 1 and Theme 2 establish their own consonant rhythmic/metric regions.

---

[21] This notion of rhythmic/metric consonance is indebted to Cohn's discussion of metric complexes in his "Dramatization" article. However, while Cohn uses an exponential relationship to define "pure duple" and "pure triple" metric complexes, his analysis examines recursion only on metric and hypermetric layers, and does not take into account tactus subdivision.

Other Relations between Subtactus and Metric Layers: *Symphonie pour cordes*

A passage from Honegger's *Symphonie pour cordes* (Honegger's second symphony) shows an intriguing relationship between subtactus and metric layers. In the first movement, rhythmic/metric consonance is maintained until m. 61: the rhythmic/metric pair is (2,4)(4). Beginning at m. 61 the number of triplet subdivisions gradually increases. From mm. 61-69, six beats express triplet subdivisions. Example 2.11 shows the increased frequency of triplet subdivisions that occur between mm. 78-96. Example 2.11a contains the score, and Example 11b provides a rhythmic reduction containing only these triple subdivisions. In mm. 78-85, three beats express triplet subdivisions (at mm. 78, 81, 83); in mm. 86-89, four beats contain triplet subdivisions (one in each measure); in mm. 90-93, eight beats (stated in counterpoint between the outer voices); and in mm. 94-96 eleven beats. In fact, in mm. 94-96, the metric space becomes saturated by triplet subdivisions on every beat save the first.

For these measures from 94-96, not only do the triplet subdivisions predominate, but the notated 4/4 meter is challenged by the first violins, whose motivic groupings express a triple metric layer of 3 quarter notes, emphasized by the accented A^6. Basses and cellos also state another triple layer in imitation, appearing one beat later than the first violins.[22] The triple level of rhythmic subdivision—the increasing presence of eighth-note triplets between mm. 78-96—then projects onto the metric level, creating an expressed layer of triple meter between mm. 94-96.

Mm. 94-96 form an important dramatic moment in the movement, operating as a flash point for the accumulated rhythmic and metric strategies. These three measures mark the point of furthest departure from the duple rhythmic/metric consonance that existed prior to m. 61. The ever-increasing triple subdivisions, heard between mm. 78-96, become promoted to the metric layer at mm. 94-96, and now the triple organization seems to wrest control from the duple organization on both layers of rhythmic subdivision and meter

Yet it is incorrect to state that the triple organization in both layers completely takes over between mm. 94-96. At the same time, these measures also retain vestiges of duple identity on the layers of rhythmic subdivision and meter. The violas and violins 2 (the lower parts of the violin *divisi*) assert both

---

22 In *Fantasy Pieces*, Krebs refers to this type of metric overlapping as "displacement dissonance." In his article "Some Extensions," Krebs originally called this "Type B Metrical Dissonance."

Example 2.11a: *Symphonie pour cordes*, score (mm. 78-96)

m. 78

m. 81

Example 2.11a (Continued)

Example 2.11a (Continued)

Example 2.11a (Continued)

Example 2.11b: Rhythmic reduction (mm. 78-96)

and triple subdivisions (see Example 2.11a). In addition, their motivic groupings square with the notated meter, maintaining the duple meter against the expressed triple meter of violins 1, cellos, and basses. For all voices, then, a duple/triple conflict is overtly stated on the two layers of rhythmic subdivision and meter. The rhythmic/metric pair for mm. 94-96 is (2,3,4)(3,4).

Determining Rhythmic Dissonance

The previous analysis shows that the gradually increasing presence of triple subdivisions at mm. 86-96 prepares the emergence of the expressed triple meter at mm. 94-96 of the *Symphonie pour cordes*. Here Honegger creates a correspondence that bridges two layers of rhythmic structure, pulse subdivision and of meter. In addition, as the triple subdivisions proliferate, they create an increasingly higher degree of conflict with the duple subdivisions that also are stated during these measures. It is possible to trace the increasing degrees of 3 : 2 conflict by comparing the duple subdivisions with the triple subdivisions in a series of stages, showing the overall process of increasing dissonance. The following develops a model that establishes and quantifies the degree of relative consonance and dissonance as a result of these conflicting subdivisions.

In order to quantify the degree of dissonance, this model can be developed in two stages: the first stage measures the degree of saturation of duple and triple subdivisions; the second stage compares these values to yield the degree of dissonance.

For three successive sections between mm. 86-96 of the *Symphonie pour cordes*, Table 2.1 gives the number of beats that express either duple or triple subdivision (sd), the total number of beats within each passage (total), and their proportional relationship (/).

Table 2.1: Saturation of duple and triple subdivisions in *Symphonie pour cordes*, mm. 86-96

Measure	Duple sd	Total	/	Triple sd	Total	/
86-89	16	16	1	4	16	.25
90-93	11	16	.69	8	16	.50
94-96	12	12	1	11	12	.92

The quotient that expresses a degree of saturation of a rhythmic subdivision (/) will be between the values of zero and one. A lower number expresses a lower degree of saturation, and a higher number expresses a higher degree. For example, duple subdivisions are stated during each beat in mm. 86-89. Therefore this maximum saturation of duple subdivisions is represented by the quotient of 1. The increased degree of triplet subdivision, indicated in the farthest column on the right, here becomes explicit, moving from .25 (mm. 86-89), .50 (mm. 90-93), to .92 (mm. 94-96).

Given the degree of saturation of duple and triple subdivisions, these then are compared to each other to determine a value for relative rhythmic consonance or dissonance. By dividing the smaller by the larger, the value again will always be between zero and one, with zero representing complete rhythmic consonance and one representing complete rhythmic dissonance. Table 2.2 demonstrates the increasing degree of 3 : 2 dissonance during this passage: .25 to .72 to .92.

Table 2.2: *Symphonie pour cordes*, Consonance/dissonance value

Measure	Duple /	Triple /	Cons/diss value
86-89	1	.25	.25
90-93	.69	.50	.72
94-96	1	.92	.92

Thus, the consonance/dissonance value illustrates an overall process that takes place in these measures, in which the move to higher degrees of dissonance culminates in the passage at mm. 94-96. It is in these three measures, too, where the duple/triple conflict then becomes expressed at a higher layer of structure, and where both 4/4 and 3/4 meter are stated.[23]

The observation of the passage through degrees of relative consonance/dissonance show the underlying dramatic progress worked out in these measures. For Honegger's *Symphonie pour cordes*, these consonance/dissonance values allow for detailed and careful analytical observations of the rhythmic structure of the compositions, and reveal the process of rhythmic conflict that is worked out within the composition.

Conclusion

In Honegger's compositions, rhythm and meter play an important constructive role. In the case of compositions such as *Mouvement symphonique (Pacific 231)*, *Sonate pour violon et piano en ré mineur*, and *Prélude pour "La Tempête"*, the composer develops rhythmic and metric strategies that serve global compositional aims. In addition, the analyses of *Sonate pour violon et piano en ré mineur*, *Prélude pour "La Tempête"*, and the *Symphonie pour cordes* show that rhythmic and metric conflicts serve dramatic and developmental processes. In these compositions, Honegger establishes conflicts and creates relationships that play out across different layers of rhythmic structure, including pulse subdivision, meter, and polymeter. Far from exhibiting a problematic mechanical approach to rhythm, Honegger's music shows careful and nuanced strategies from the lowest to the highest levels of compositional organization.

---

[23] It is also possible to design this statistical model to address degrees of dissonance on the metric layer as well. Mm. 94-96 provide a ready example in which both 3/4 and 4/4 meters are expressed throughout, creating maximum dissonance at this layer of structure. Further refinements could address levels of dissonance at *both* the subtactus and the metric layers.

Chapter 3

Contrapuntal Structures and Symmetries

Introduction

This chapter begins with a discussion of Honegger's extensive student counterpoint studies at the Paris Conservatoire, then moves on to examine the role of counterpoint in Honegger's compositions. Counterpoint provided Honegger with a consistent organizational foundation that operated beneath an eclectic approach to pitch organization. The notion of counterpoint here reflects not only the use of contrapuntal devices such as fugue, stretto, canon, and imitation, but also suggests the presence of a contrapuntal ideal grounded in contrary motion principles.

These contrary motion principles operate as contour symmetries that may control the foreground verticalities, and three types of symmetries are presented in order of increasing specificity. Following, the chapter presents a discussion of two types of abstract symmetries, pitch-class inversion and the interval cycle. These abstract symmetries, used in the first of Honegger's *Trois contrepoints*, organize not only the vertical but the horizontal dimension, and are capable of organizing longer linear spans. Finally, the analyses regard imitative and fugal passages from several works, including a student fugue, the unpublished Sonata for Violin in D minor, and the Sonatina for Clarinet, showing the relationship of the horizontal to the vertical dimensions, and illustrating the relationship of contrapuntal and metric structure.

Years of Apprenticeship

Honegger's attraction to counterpoint and contrapuntal procedures was lifelong. As early as 1920 he wrote of his "perhaps exaggerated" tendency to seek polyphonic complexity, citing Bach as his compositional model.[1] Dismissing any programmatic associations in his most famous composition, *Mouvement symphonique (Pacific 231)*, he described it instead as a "diversified chorale,

---

[1] Arthur Honegger, *Ecrits*, Textes réunis et annotés par Huguette Calmel (Paris: Librairie Honoré Champion, 1992), 34.

strewn with counterpoint in the manner of J.S. Bach."[2] Moreover, a number of pieces are overtly based upon specific contrapuntal procedures, such as the fugues and fughettas from *Prelude, Fugue, and Postlude (d'après Amphion)*, Two pieces for organ, and the *Prélude, Arioso, Fughette (sur le nom de Bach)*. In addition, several compositions embed fugal sections within the movement of a sonata, such as the Second Violin Sonata and Sonatina for Clarinet.

The appeal and fascination with the discipline of counterpoint may be traced even to the earliest works. For example, a work of juvenilia, the *Six Sonates pour pianoforte et violon* (1908)[3] contains two sonatas whose final movements are labeled fugues. They are, however, not fugues but essays in invertible counterpoint, in which each of three melodic lines is introduced consecutively in the three parts: violin, piano right hand, and piano left hand. The lines are then circulated, untransposed, into a different part. Figure 3.1, which displays the arrangement of voices for the last movement of the fourth sonata, indicates a highly systematic contrapuntal organization for the voice relationships.

Figure 3.1: Last movement, Sonata no. 4 from *Six Sonates pour pianoforte et violon*
S= Subject; 1= Countersubject 1; 2 = Countersubject 2

Violin			S	1	2	S
Piano (RH)		S	1	2	S	1
Piano (LH)	S	1	2	S	1	2
Measure	1	9	17	26	35	43

The seven years of counterpoint training at the Paris Conservatoire (1911-1918) supplied a rigorous technical and compositional foundation for the development of Honegger's musical thought. During these years he studied with André Gedalge, the author of *Traité de la Fugue*,[4] who had also taught Maurice Ravel, Florent Schmitt, and Darius Milhaud. Honegger twice attempted the *concours de contrepoint*, receiving only a *deuxième accessit* in his final year at the Conservatoire. He captured the rivalry between the two counterpoint classes in a letter to his parents:

> There are two counterpoint classes at the Conservatory (Gedalge's and Caussade's) and two composition classes (Widor's and Vidal's). Caussade's is the pedantic class par excellence, so most of the prize

[2] Arthur Honegger, *I am a Composer*, trans. by Wilson Clough (New York: St. Martin's Press, 1966), 101.

[3] Le Havre, 1908. Unpublished, contained at the Bibliothèque Nationale.

[4] André Gédalge, *Treatise on the Fugue*, trans. and ed. by Ferdinand Davis (Norman, OK: University of Oklahoma Press, 1965).

> winners come from among his students. But no musician has ever come out of it, for the simple reason that Caussade is very strong on the technical front but completely empty on the artistic one.[5]

In his letter, Honegger also went on to criticize the harmonically-based counterpoint that characterized contrapuntal pedagogy in German-speaking institutions such as the conservatories at Vienna and Zürich (the latter where he had studied from 1909-11).

Honegger's counterpoint studies with Gedalge, a significant portion of which are now at the Paul Sacher Foundation in Basel, are primarily in eighteenth-century style, although the studies also contain examples of sixteenth-century contrapuntal techniques. Over 300 pages, they include some 52 chorale harmonizations, 188 cantus firmus exercises, 10 studies in invertible counterpoint, 51 exercises in three-voice species counterpoint, and some 40 partial or complete fugues. The contrapuntal study culminated in the fugues, which, like Gedalge's treatise, host a number of technical fugal devices, including stretto, subject in augmentation and diminution, and subject in contrary motion.

Geoffrey Spratt describes the evolution of Honegger's student compositions as a gradual replacement of harmonic vertical organization with contrary-motion principles.[6] Spratt regards Honegger's earliest use of contrapuntal techniques in his compositions as problematic, since these works lack a reconciliation between the harmonic and contrapuntal dimensions, a reconciliation eventually achieved by the compositions of 1917. By this point, linear contrapuntal procedures provide the underpinning for Honegger's harmonic vocabulary, and his pitch language parallels—and is perhaps indebted to—Schoenberg's prewar works such as *Pierrot Lunaire*. In addition, these compositions make extensive use of imitative textures, fugue, canon, and techniques such as melodic augmentation, diminution, and inversion.

Significantly, then, these traditional techniques and procedures of counterpoint such as outer voice relationships, principles of invertible counterpoint, imitative textures, and imitative procedures such as fugue—techniques that were developed during the years of study at the Conservatoire—gradually inform Honegger's "free" compositions. Yet, while these compositions exhibit these techniques and procedures, they are adapted to harmonic and pitch languages far outside the strictures of Baroque tonality or Renaissance modality that are present in his student counterpoint studies. Once absorbed, the

---

[5] Harry Halbreich, *Arthur Honegger*, trans. Roger Nichols (Portland, Oregon: Amadeus Press, 1999), 31.

[6] Spratt's discussion of contrapuntal procedures and contrary motion organization occurs throughout the book; for its early use and concomitant shift away from additive harmony, see Geoffrey Spratt, *The Music of Arthur Honegger* (Cork: Cork University Press, 1987), 14, 16, 22-3.

contrapuntal techniques form the basis of compositional organization for entire compositions or portions of compositions. If overstating the case, Spratt's evaluation nevertheless addresses this orientation: "The predominant stylistic feature of all Honegger's music is the use of contrapuntal polyphony."[7]

For Honegger, techniques of counterpoint provided a means of achieving compositional coherence. Even for the compositions that do not employ overt contrapuntal imitative procedures such as fugue or canon, the contrapuntal foundation controlling outer-voice relationships allows organizational consistency and compositional unity *despite* the radical differences in pitch/harmonic language and the vast disparity of musical styles and genres. Analysis of this contrapuntal foundation also provides an antidote to an analytical approach that merely lists the variety of techniques of pitch organization, such as tonality, extended tonality, and atonality.[8] While it is an overstatement to assert a contrapuntal origin for all of Honegger's music, counterpoint provides, nevertheless, an analytical perspective for a significant portion of it.[9]

Vertical Organization: Outer Voice Relationships

What are the organizing factors that control the foreground verticalities for the music of Honegger? Certainly its contrapuntal foundation would seem to call for

---

[7] Ibid., 212.

[8] This is essentially the approach taken in the chapter "Honeggers Stil," in Pierre Meylan, *Arthur Honegger: Humanitäre Botschaft der Musik* (Frauenfeld and Stuttgart: Verlag Huber, 1970), 189-216.

[9] Historically, writings on counterpoint provide a pedagogical basis for composition by defining a series of restraints that control the vertical and, occasionally, the horizontal dimensions of the musical texture. Regulation and control of the *vertical* dimension forms the foundation of counterpoint—as its etymology suggests—and the historical evolution of contrapuntal pedagogy and thought is distinguished by additions to permissible vertical intervals and, often, increased freedom in their use. For example, the contrapuntal treatises for the *seconda prattica* were characterized by freer dissonance treatment in contrast to those treatises of the *stilo antico*. See Christoph Bernhard, *Tractatus compositionis augmentatus* (c. 1650); Guillaume Nivers, *Traité de la composition musicale* (Paris: 1667); Jan Adams Reincken, *Kompositionslehre* (c. 1670); Giovanni Bononcini, *Il Musico prattico* (Bologna: 1673); Angelo Berardi, *Documenti armonici* (Bologna: 1681).

Beginning with Ernst Kurth's 1917 *Grundlagen des linearen Kontrapunkts*, many discussions of counterpoint in the twentieth-century emphasize the linear dimension, presenting the resultant verticals as byproduct. See *Ernst Kurth: Selected Writings*, ed. and trans. by Lee Rothfarb (Cambridge: Cambridge University Press, 1991), especially 39-40. This focus is partly a result of the absence of *a priori* distinctions between vertical consonance and dissonance within the corpus of twentieth-century music. This axial shift from a vertical to a horizontal perspective in contrapuntal thought is evinced by one analyst's term *rücksichtsloser Kontrapunkt* (reckless counterpoint), used to describe contrapuntal music of Stravinsky and Hindemith that seems to disregard the resultant vertical intervals formed by simultaneous melodic lines. See Willi Apel, ed. *Harvard Dictionary of Music*, 2d ed. (Cambridge, Mass.: Harvard University Press, 1977), s.v. "Linear counterpoint."

an analytical system that takes into account structured voice relationships. On the one hand, traditional contrapuntal theory, founded upon the structuring of voice relationships, is of limited use, since the circumscribed set of permissable pitch or pitch-class intervals limits its applicability for music not exclusively modal or tonal. On the other hand, the theory of pitch-class sets provides a powerful tool for analysis of a vast repertoire of vertical sonorities, but in considering simultaneities as *unordered* collections of pitch classes, it provides little information on their contrapuntal origin. Therefore such an approach does not necessarily address the structuring of relationships between voices.

Instead, as a point of departure, we may consider the notion of inversional symmetries which structure the simultaneities on the musical surface. For Honegger, symmetry provided a method for systematically organizing voice relationships despite differences in musical syntax.

Inversional symmetry will be developed here in three stages of increasing specificity. The first and most general type is contour symmetry, which occurs when direction in a single voice is complemented by opposing direction in another voice. Example 3.1a provides one example. Note that the interval distance need not be identical in both voices; only the overall contour may be mirrored in each line. A second type of inversional symmetry is generic interval symmetry. Example 3.1b shows that this takes place through the preservation of generic—usually diatonic—intervals in opposite directions. Generic interval symmetry is traditionally a tonality-preserving feature in which an interval is mirrored by its generic equivalent. For example, an ascending major second may be potentially mirrored by a major or minor second. The third and most specific form of inversional symmetry is specific interval symmetry, which preserves exact intervalic distance in contrary directions. Example 3.1c provides an example of specific interval symmetry.

Example 3.1: Inversional symmetries

1. Contour Symmetries[10]

Analytical models for melodic contour provide a means for representing melodic direction outside of the specific pitch and pitch-class context. Methodology for study of musical contour allows the modeling of direction for a single melody, and ways to evaluate its relationship to other melodies. Application of this method to outer-voice relationships reveals the contour symmetry of simultaneous melodies.

Example 3.2, which contains the primary motive that forms the opening of Honegger's atonal opera *Antigone* (1927), shows a contour symmetrical relationship. Except for the exchange of C/Db in the two voices across the barline, this contour symmetry is independent of pitch-specific symmetry. In this example, the contour symmetries are not used to preserve tonality, but instead articulate surface voice relationships within a nontonal idiom.

Example 3.2: Contour symmetry in *Antigone*

In contrast to the atonal environment of Example 3.2, the overt diatonicism of a passage from Rehearsal 25 of Honegger's *Mouvement Symphonique #2 (Rugby)* supports several contour symmetries. Honegger's piano reduction for the passage appears in Example 3.3a. Example 3.3b isolates these symmetries. At the level of the quarter note, the passage is initiated with the simultaneous ascent and descent of the triplet figure stated in stretto. Ex. 3.3b, no. 1 separates out these two lines, showing more clearly the contour symmetry. Additionally, the opening move of the upper parts (lines C and A) in stretto at the level of the beat loosely asserts a contour symmetry: in Example 3.3b, no. 2, the contour segment of <+ - -> in the lower voice, second beat, is stated simultaneously against <-> in the upper voice.

In this passage, contour symmetry is not only stated in the vertical dimension, but also projected into the linear dimension in the same passage. At

10 A few music-theoretic studies have developed analytical models for musical contour, including Michael Friedmann, "A Methodology for the Discussion of Contour: Its Application to Schoenberg's Music," *Journal of Music Theory* 29 (1985): 223-48; Elizabeth Marvin and Paul Laprade, "Relating Musical Contours: Extensions of a Theory for Contour," *Journal of Music Theory* 31 (1987): 225-67; and Robert Morris, "New Directions in the Theory and Analysis of Musical Contour," *Music Theory Spectrum* 15/2 (Fall 1993): 205-28. Discussions of contour symmetry here will use the symbol "+" to refer to ascending contour, and "-" to refer to descending contour.

the two-measure level, the bass motion asserts contour symmetry with its pattern of ascending and descending scales: in Example 3.3b, no. 3 the contour pattern of the first two measures <+ + + + + + +> is matched in the third

Example 3.3a: *Mouvement symphonique #2 (Rugby)*, Rehearsal 25

Example 3.3a (Continued)

Example 3.3b: Contour symmetries

1.

2.

3.

4.

parts reflects the axis of symmetry on D, the focus of pitch centricity. The passage becomes:

Triadic root of upper part: (1 0 1 0 2 1 0 6 5 4 5)
Triadic root of lower part: (1 2 1 2 0 1 2 3 4 5 4)

Each of the note pairs sum to two (modulo 7). The axis of generic interval symmetry is half of this sum: 1 or, in this instance, pitch class D.

3. Specific Interval Symmetries

Inversional symmetry based upon specific intervals provides the focus for discussing compositional structure in a number of works by twentieth-century composers, including Bartók, Berg, Varèse, and others.[13] This species of symmetry provides a model for revealing musical organization at the foreground, as well as at potentially larger levels within a composition. As noted here with other types of symmetry, specific interval symmetry may be deployed in the horizontal as well as vertical dimension. For example, the melody in m. 29 of the second movement of the *Symphonie pour cordes* shows its operation in the linear dimension. Example 3.5 shows that the intervalic relationship of the viola and cello melody is identical backward as well as forward. In the example, the ascending pattern of semitones is <4 1 1 4>, and the melody is inversionally symmetrical around the central pitch F.[14]

[13] From as early as 1955, the music of Bartók has attracted a significant amount of attention in this area; see Elliott Antokoletz, *The Music of Béla Bartók: A Study of Tonality and Progression in Twentieth-Century Music* (Berkeley and Los Angeles: University of California Press, 1984); Jonathan Bernard, "Space and Symmetry in Bartók," *Journal of Music Theory* 30/2 (Fall 1986): 185-201; Wallace Berry, "Symmetrical Interval Sets and Derivative Pitch Materials in Bartók's String Quartet No. 3," *Perspectives of New Music* 18 (1979-1980): 287-379; Richard Cohn, "Inversional Symmetry and Transpositional Combination in Bartók," *Music Theory Spectrum* 10 (1988): 19-42; Colin Mason, "An Essay in Analysis: Tonality, Symmetry, and Latent Serialism in Bartók's Fourth Quartet," *Music Review* 18 (1957): 189-201; George Perle, "Symmetrical Formations in the String Quartets of Béla Bartók," *Music Review* 16 (1955): 300-312; Leo Treitler, "Harmonic Procedures in the Fourth Quartet of Béla Bartók," *Journal of Music Theory* 3 (1959): 292-8. Elsewhere George Perle has developed the notion of inversional symmetry in *Twelve-Tone Tonality* (Berkeley and Los Angeles: University of California Press, 1977), and "Berg's Master Array of the Interval Cycles," *Musical Quarterly* 63/1 (1977): 1-30. Dave Headlam develops Perle's notions of symmetry for Berg's music in *The Music of Alban Berg* (New Haven and London: Yale University Press, 1996). On inversional symmetry in the music of Schoenberg, see David Lewin, "Inversional Balance as an Organizing Force in Schoenberg's Music and Thought," *Perspectives of New Music* 6/2 (1968): 1-21. Inversional symmetry in pitch space provides the central thesis for Jonathan Bernard, *The Music of Edgar Varèse* (New Haven: Yale University Press, 1987).

[14] The ostinato accompanimental figure manifests other symmetries in the same measure. The upper stave of the *divisi* violins assert a contour symmetry, while the two lower

Example 3.5: Specific interval symmetry, *Symphonie pour cordes*, ii (m. 29)

In contrast to the atonal environment of Example 3.5, specific interval symmetry is likewise used to embellish a tonal harmonic environment. Example 3.6a shows the opening to the second movement from *Le Roi David*. Here the presence of an E major triad (in first inversion) and its dominant (in third inversion) implies a harmonic context of E major. Note that in the first ten measures, the voice motion in the center staff articulates specific pitch symmetry, creating a voice-exchange articulated on the downbeat of each measure: the C#/B moves to C/C, then to B/C#. The voice exchange elaborates a tonal language that includes the use of added-chord harmony, with the C# operating as added sixth above the tonic, and added ninth above the dominant.

Example 3.6b shows the symmetry of these ten measures along with its continuation at the second phrase at the following three measures. While the second phrase maintains specific interval symmetry, the focus of the voice exchange shifts. Instead of embellishing B/C#, the upper part is shifted up a half step, embellishing instead B/D from m. 11 ff. The elaborated voice exchange now occupies two measures, elaborating a v^9 harmony within the key of E major; again maintaining and embellishing a tonal harmonic environment.

The principle of specific interval symmetry is often exemplified through pitch dyads around an inversional center, in which the music may assert or imply a single axis of inversion. Given the numerical representation of pitches (i.e., C = 0, C# = 1 ... B = 11), the summing of the pitches will yield an axis of inversion around a single pitch when the dyads sum to an even number, or two

parts of the violins state specific interval symmetry, with the violins 2 stating B-G# against the C-A in the lower line of violins 1.

pitches a half-step apart when the dyads sum to an odd number.[15] This axis of inversion based upon the structured voice relationships need not be presented overtly.

In addition to describing an abstract precompositional structure, specific pitch symmetry may serve as a focus of directed musical motion. For the analysis of twentieth-century music, then, specific interval symmetry provides a degree of analytical potency absent from pitch-class set analysis. Models of inversional

Example 3.6a: "Song of David, the Shepherd," from *Le Roi David*

[15] Milton Babbitt introduced the notion of "index number" in "Twelve Tone Rhythmic Structure and the Electronic Medium," *Perspectives in New Music* 1/1 (1962): 49-79.

Example 3.6a (Continued)

Example 3.6b: Specific interval symmetry

pitch symmetry may include a linear and teleological component lacking in analytical descriptions confined to pitch-class set recurrence and inclusional relationships.[16]

Specific interval symmetry may provide a sense of directed motion, effected in one of two ways. First, it may imply motion toward the inversional axis, supplying the gravity that impels the pitch content toward arrival on this axis. Here, this motion to the inversional axis brings about literal pitch centricity within a passage or a composition. Second, this model provides a concept of directed motion through the completion or fulfillment of inversional symmetry with the constituent dyadic pairs. In this case, the withholding or absence of an inversional partner from a dyadic pair requires its eventual presence. The directed linear motion arises from the implication, or the eventual completion, brought about through the arrival upon the missing half of the inversional pair.[17]

[16] The absence of directed motion in pitch-class set analysis provides the impetus for William Benjamin's critique in "Ideas of Order in Motivic Music," *Music Theory Spectrum* 1 (1979): 23-34.

[17] The concept of implied directed motion through inversional symmetry is explored in Bruce Archibald, "Some Thoughts on Symmetry in Early Webern: Op. 5, No. 2," *Perspectives of New Music* 10/2 (Spring/Summer 1972): 159-63; David Lewin anthropomorphizes the concept in discussing "generative lusts" implied by inversional symmetry in "Transformational Techniques in Atonal and Other Music Theories," *Perspectives of New Music* 21/1-2 (Fall/Winter 1982 and Spring/Summer 1983): 312-71. Some of the analysis of the *Symphonie pour cordes* in Chapter 5 will further explore this issue of directed motion through inversional symmetry.

Abstract Symmetries: Pitch-class Inversional Symmetry and Interval Cycles

Pitch-class Inversional Symmetry

Specific interval symmetry refers to symmetry in pitch space. That is, this symmetry *necessarily* takes into account the specific registral placement of each pitch. In the case of specific interval symmetry, dyadic pairs will have a single axis of inversion. Several writers, most notably George Perle and Elliott Antokoletz have advanced the more abstract principle of symmetrical inversion in pitch-class space.[18] Here, symmetry is represented through pitch-class dyads which are given numerical representation (where C = 0, C# = 1, etc.) and summed. Inversional symmetry is maintained when the dyadic pairs sum both to the same integer *modulo* 12, or to an integer at a distance of (+ or -) 6. The inversional dyads now may support two abstract axes of inversion at the distance of a tritone.

In contrast to symmetrical inversion in pitch space, symmetrical inversion in pitch-class space removes the necessity of fixed registral placement. Thus the generalization from pitch to pitch-class inversional symmetry potentially abandons the contrary motion principles present in our three earlier models of contour symmetry, general interval symmetry, and specific interval symmetry. As a result, pitch-class inversional symmetry loses somewhat the analytical power to explain *directed* motion underlying surface voice relationships. Moreover, pitch-class symmetry replaces the single fixed axis of symmetry by two abstract pitch-class axes.

Interval Cycles[19]

Perle and Antokoletz present inversional symmetry through the construct of symmetrical cells. This construct is then generalized to symmetry based upon equal subdivisions of the octave provided by a single generating interval class (ic). The resultant partition of the musical space is called an interval cycle. The complete set of interval cycles consists of one cycle of ic 1, two cycles of ic2 (the

---

[18] See especially Antokoletz, *Bartók*; and Perle, *Twelve-Tone Tonality*, "Symmetrical Formations," and "Berg's Master Array."

[19] Musical organization founded upon this principle of symmetry is discussed in Antokoletz, *Bartók*, "Interval Cycles in Stravinsky's Early Ballets," *Journal of the American Musicological Society* 39/3 (Fall 1986): 578-614; and "Principles of Pitch Organization in Bartók's Fourth String Quartet," *In Theory Only* 3/6 (September 1977): 3-22; Philip Lambert, "Interval Cycles as Compositional Resources in the Music of Charles Ives," *Music Theory Spectrum* 12/1 (1990): 43-83; Ann McNamee, "Bitonality, Mode, and Interval in the Music of Karol Szymanowski," *Journal of Music Theory* 29/1 (1985): 61-84; Charles Porter, "Interval Cycles in Alban Berg's String Quartet Opus 3," *Theory and Practice* 14/15 (1989-1990): 139-78; and Dave Headlam, *The Music of Alban Berg*. A number of George Perle's essays which treat this topic have been collected in *The Right Notes* (Stuyvesant, New York: Pendragon Press, 1995).

two whole-tone scales), three cycles of ic3 (the three diminished-seventh chords), four cycles of ic4 (the four augmented triads), one of ic5, and six of ic6. Perle notes:

> The interval cycle is a means of symmetrically partitioning, and thus imposing an ordering upon, the functionally undifferentiated pitch classes of the twelve-tone scale. It is a concept that evolves ... concomitantly with the weakening and eventual elimination of traditional harmonic functions and the replacement of a diatonic scale of unequal degrees by an undifferentiated semitonal scale.[20]

Interval cycles form a background repository for structuring foreground harmonic and melodic events, and may control longer-range motion through the transpositional cycles of an ic.

The following analysis shows the use of both pitch symmetry and pitch-class symmetry, and draws upon the notion of the interval cycle in order to show how these elements participate in both vertical and linear design.

Vertical and Linear Structures in *Trois contrepoints*

The notion of specific interval symmetry is in evidence in an excerpt from the first of the *Trois contrepoints* (1922). In the excerpt, moreover, the axis of inversion plays a privileged role in structuring the pitch centricity and formal recapitulation of the movement. The passage appears in Example 3.7a.

The formal return of the opening material occurs in m. 37. Beginning in m. 34, a chromatic descent in the cello is paired with a melodic ascent articulated on downbeats in the upper part. The major second (E4-F#4) straddling the mm. 34-35 barline in the oboe melody takes place simultaneously with the descent from C4 to A# in the cello, which chromatically fills in this motion. Each beat through the next measure asserts a chromatic wedge with D4 providing the literal inversional axis. Example 3.7b simplifies the texture to show this. However, 3.7b also shows that an octave shift downward in the bass in m. 36 alters this literal inversional axis, shifting it to Ab3 away from D4. Yet clearly the pitch arrival on D in both voices in m. 37, corresponding to the formal recapitulation, contradicts the notion of a shift of pitch centricity to Ab. In one sense, this octave displacement in the bass may correspond to the octave displacement in the oboe at the m. 37 arrival upon D, reorienting the pitch axis of symmetry back to D, albeit now moved from D4 to D3. In another sense, the octave displacement in the oboe at the m. 37 arrival upon D, reorienting the pitch axis of symmetry

---

[20] From Perle's essay "Scriabin's Self-Analyses," in *The Right Notes*, 23.

Example 3.7a: Mm. 34-37 from #1 of *Trois contrepoints*

Example 3.7b: Axis of inversion

Example 3.7c: Secondary wedge

back to D, albeit now moved from D4 to D3. In another sense, the octave displacement in the bass nevertheless extends inversional symmetry here to the pitch-class dimension where the pitch-class dyads that sum to 4 and 10 still highlight the arrival of D as the focus of the directed motion. Before looking at this issue further, it is necessary to investigate the passage in greater detail.

The excerpt contains another level of inversional symmetry. The compound melody in the upper voice twice creates secondary inversional wedges, initiated both times on the metric downbeats of mm. 35-36: the first initiated with F#4, the second with A#. Example 3.7c isolates these secondary wedges. Thus, the passage supports simultaneous axes of inversion, the principal axes shown in Example 3.7b, and the secondary ones given in Example 3.7c. The axis of the principal inversional wedge serves to prepare a point of arrival upon D at m. 37, while the F# and A# axes of Example 3.7c are subordinate to this directed motion.

Other factors contribute to heighten the arrival on D at m. 37. Once the wedge motion is initiated from the D^4 in mm. 34, each pitch class except for D is represented in these measures. Additionally, the final beat of m. 36 participates in establishing a tonal gravity, where the tritone G and C# forms a third and augmented sixth above the bass Eb, generating a dominant sonority a half-step above the point of arrival.

In addition to the foreground factors that contribute to the arrival upon D, other elements structure this arrival in a longer-range sense. The arrival at D is brought about through completion of two different interval cycles in the two separate instrumental lines.

Honegger configures the oboe part to ascend on downbeats through an ic4 cycle: m. 34 begins with D, moves to F# at the downbeat of m. 35, and ascends to A# at the downbeat of m. 36. The oboe completes the ic4 cycle with the arrival of D at the downbeat of m. 37. These metric downbeats are underlined further through the durational accent given to F# and A#. Moreover, as we have seen in Example 3.7c, these pitches form the begin points of the inversionally symmetrical motions: D launches the principal outer voice inversional wedge, while F# and A# initiate the upper voice secondary motions. The upper voice is structured through a completed ic4 cycle moving D-F#-A#-D. To an extent, the ic4 cycle in the cello on these metric downbeats, progressing from A# (m. 35), to F# (m. 36) before arriving on D (m. 37), mirrors the ic 4 cycle in the oboe.

As noted above (and shown in Example 3.7b), the octave displacement in the cello at m. 36 discards the literal center of inversion, giving way to a pitch-class and abstract center of inversion in order to maintain D as its axis. Yet other considerations explain this move: where the completion of interval class 4 governs the motion in the upper part, this m. 36 octave displacement completes a different, and longer-range, interval cycle in the lower part.

Example 3.8a begins at m. 29, seven measures before the formal recapitulation. Note that while the two-measure oboe melody begins identically at m. 30 and at m. 32, the cello twice chromatically fills out a minor third span: first progressing from D-B, then B-G#. In addition, there is an alteration to the two-measure oboe melody. This change conforms to the T3 relationship, and the oboe melody, which first terminates with F4 at the downbeat of m. 32, ends with D4 at the downbeat of m. 34. In the cello, the ic3 cycle (begun with D, moving to B, then G#) is prevented from completion once the mm. 34-36 chromatic wedge motion begins.

The completion of this ic3 cycle in the cello seems to be thwarted by the chromatic wedge interruption. Nevertheless, as Example 3.8b indicates, the octave displacement in the cello with its final three pitches F^2, E, Eb, and D connects these pitches back to the previous ic3 cycle motion. Now stated in augmentation, Honegger brings out the association with the earlier ic3 motion of D, B, and G# by the tenuto markings for the cello at m. 36. Hence the cello's registral shift of m. 36, negating the D^4 axis of inversion, nonetheless serves to complete the ic3 cycle begun seven measures earlier.

Thus, Example 3.8b shows that both voices articulate a conflicting interval cycle in the linear dimension. The upper voice is metrically organized through the completion of an ascending ic4 cycle, while the pitch organization of the cello is somewhat more complicated. On the one hand, the metric organization of the cello in mm. 35-37 participates with the oboe in stating the ic4 cycle; on the other, the longer-range motion and registral shift articulates a descending ic3 cycle between mm. 31-37. This cyclic descent of ic3, although interrupted, nevertheless is completed simultaneously with the upper voice.

In the final two measures of the excerpt, principles of inversional symmetry govern the verticalities on the musical surface. These first use specific interval (pitch) symmetry, then yield to pitch-class inversional symmetry. The longer-range structure emerges through the completion of a conflicting interval cycle in each voice. The beamed half notes of Example 3.8b shows that Honegger makes this conflict of interval cycle particularly audible through contour symmetry: note that the ascending ic4 cycle is mirrored by the descending ic3 cycle.[21]

[21] As Example 3.8b indicates, the completion of the ascending ic4 cycle upon D^5 takes place on the second beat.

Example 3.8a: Mm. 29-37 of *Trois contrepoints*

Example 3.8b: Conflict of interval cycle, mm. 31-37

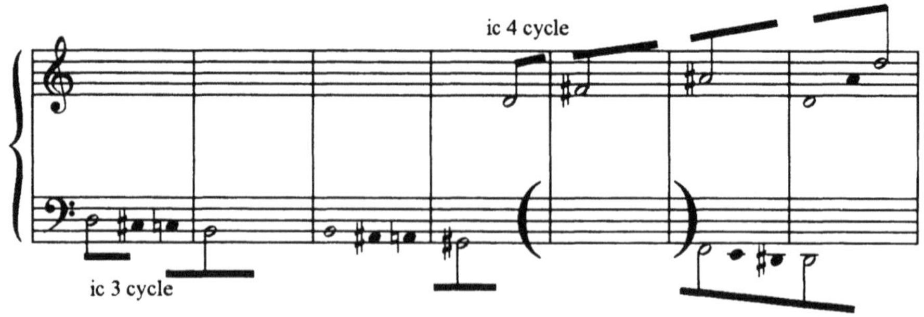

Linear Contrapuntal Structures: Correspondences Between Vertical and Linear Organization

The previous analysis shows how abstract symmetries of pitch-class inversional symmetry and interval class control the organization of musical spans that extend beyond foreground verticalities. The interval cycles ic3 and ic4, shown in Example 3.8b, organize the linear dimension, and the completion of these interval cycles provides a middleground organization to the passage. In addition, other factors may contribute to linear organization, and we can examine some works for which the vertical dimension determines the horizontal organization. In these, the foreground verticalities determine, and provide a rationale for, transpositional cycles that project and elaborate the musical structure.

Among Honegger's student contrapuntal studies, an unusual and interesting example is seen in Example 3.9a, which contains a section of a fugue, and involves several technical devices, such as subject in contrary motion, stretto, and pedal point.[22] The excerpt is interesting in its treatment of pedal point, which begins in the third measure of the excerpt. One unusual aspect of the two-measure motive begun in the alto, bracketed and labelled "Mm. Contraire" by Honegger in the score, is its major-seventh leap from G-F# across the barline, straddling the third and fourth measure. The two-measure motive is presented in stretto with the tenor beginning in the fourth measure, overlapped at a distance of one measure.

While the overlapping series of stretti continue, indicated by Honegger's brackets, more unusual is the relationship of the material that follows. A comparison of mm. 4-5 to the following two measures, as well as to the next pair of measures, reveals that the material of each of the three upper voices returns transposed up a half-step for each successive pair of measures. The reduction in Example 3.9b shows that the three ic1 transpositions of the diminished-seventh sonority results in a change of its implied function: the implied vii^{o7}/g in m. 4 returns as vii^{o7}/Bb. Despite the shift of implied function, the three transpositions ultimately bring about the enharmonic and cyclic return of the original diminished seventh sonority at m. 10.

Here Honegger exploits the cyclic properties of the ic3 cycle. The ic3—the diminished-seventh—sonority elaborated within each measure becomes *projected* into the linear dimension through the three systematic half-step transpositions, eventually bringing about the cyclic return of the initial sonority. The passage, then, highlights the invariant properties of the ic3 vertical structure. In the sketchbook, the overheated chromatic treatment of pedal point is eventually "corrected." On the following page Honegger rewrites the fugue, and the material above the dominant pedal is rewritten and harmonically tamed.

---

22 Unpublished, contained at the Paul Sacher Foundation.

Example 3.9a: Pedal point from student fugue

Example 3.9a (Continued)

Example 3.9b: T1 scheme, mm. 4-10

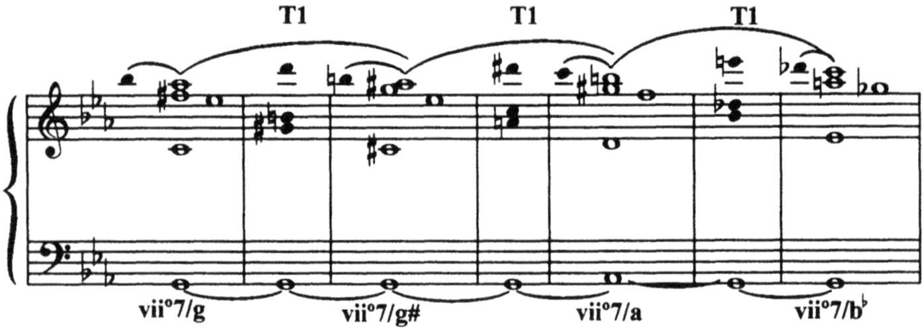

Nevertheless, the excerpt illuminates the contrapuntal logic that Honegger developed in his free compositions.

A similar contrapuntal logic is evident in a composition also written during Honegger's student years, the *Sonate pour violon et piano en ré mineur* (1912),[23] completed during Honegger's second year at the Conservatoire. Example 3.10a contains a passage from the third movement. In the passage, Honegger deploys a contrapuntal texture comprised of three-voice imitation, begun by the piano right hand on the downbeat, followed by the violin (second beat), and the piano left hand (third beat). The 3/4 meter allows the systematic playing out of this motive in the three voices at imitation at the level of the quarter note.

Additionally, this material recurs in each measure transposed down at the level of three semitones; the downbeats of the upper part of the piano and the violin assert the pitches Bb-G-E-C# between mm. 96-100. This ic3 transpositional scheme therefore projects into the linear dimension the diminished-triad verticality that is stated on each successive downbeat, and the cyclic property of the downbeat sonorities becomes expressed at each transpositional level. Moreover, the four statements of the material systematically exhaust the minor third cycle before the following sectional and textural shift.

There is yet another level of correspondence operating in the excerpt. The ic3 sonority that recurs in a network of ic3 transpositional cycles between mm. 96 -99 corresponds to the 3/4 meter. The following material, beginning at m. 100, shifts both the harmonic and the metric environment. Now the 3/4 meter yields in the piano accompaniment to a duple 2/4 meter, and, simultaneously, the new pitch material asserts an ic2 cycle (whole-tone) harmonic environment stated in both accompaniment and violin. Thus the passage creates a correspondence of

23 Unpublished, contained at the Paul Sacher Foundation.

Example 3.10a: Mm. 96-100 from Sonata in D minor

Example 3.10b: T3 scheme

interval cycle and meter. The triple meter conveys the ic3 material, expressed both vertically and horizontally, while the duple meter which follows expresses the ic2 material.

Beneath the evident imitative textures in both the student fugue and the early violin sonata, the linear contrapuntal events are governed by a systematic organization that relates the vertical dimension to the horizontal. In both works the ic3 cycle referential sonorities which are prevalent upon the surface of the music determine the levels of transposition which bring about cyclic invariance. In the fugue, the three successive statements at T1 bring about the cyclic return of invariant pitches; in the sonata, the four statements at T3 systematically exhaust the transpositional cycle. In the sonata, too, the further correspondence between interval cycle and meter heightens the sectional change within the composition.

Contrapuntal and Rhythmic Structures: Sonatina for Clarinet

Several of Honegger's early chamber works include strict fugal expositions or fughettas embedded within a larger movement, often providing a textural contrast to the surrounding music. For example, the second Violin Sonata (1919) contains within the first movement development section a three-voice fugal exposition in which the answer occurs a perfect fifth from the subject. The second theme from the first movement of the Sonatina for Clarinet (1922) likewise includes a fugue, containing a four-voice exposition of a two-measure subject. Rather than the traditional intervalic distance of a perfect fifth, the answer appears a tritone away from the subject. The fugal passage appears in Example 3.11a.

The first statement of subject and answer (mm. 1-4) states all twelve pitch classes. The two-measure subject lacks pcs D, Eb, F, and Bb, all of which are stated in the answer which follows. On the downbeat of m. 3, the left hand states Eb, D follows in the RH (second eighth note), Bb appears in the last eight note of the left hand, and F appears in m. 4 as E# on beat two of the left hand.

Despite this chromatic aggregate saturation, the specific interval symmetry, brought about by the succession of LH accompaniment pitches, plays a role in structuring the return of the following subject. Example 3.11b shows that within mm. 3-4 the first three pitch pairs (D#/B, E/Bb, A/E#) state sum 2 dyads. Here the axis of inversion of these sum 2 dyads is G. (The final accompanimental pitch of F# in m. 4 does not participate in these dyadic symmetries.) Honegger projects the axis of inversion of these dyads (pc 7, or G) upwards two octaves to initiate the restatement of the subject on G in m. 5.

In addition to the specific interval symmetries stated in the linear dimension, the texture creates vertical inversional symmetries at mm. 7-8, the last two measures of the exposition. Example 3.11c simplifies the texture, showing the outer voices of the piano, which accompany the restatement of the answer. Note that the axis of inversion shifts between mm. 7-8. The embellished ic2 ascent in the piano RH (A-B) mirrors the ic2 descent in the left hand (B-A). Leading with the upper voice, these two voices appear at a distance of an eighth note in the score. The dyadic sum of 8 for the outer parts (B/A, A/B) then shifts in the second measure to sum 9 dyads (G#/C#, G/D). Again the upper voice

Example 3.11a: Sonatina for Clarinet

Example 3.11a (Continued)

Example 3.11b: Inversional dyads around G, mm. 3-5

Example 3.11c: Inversional symmetry at mm. 7-8

leads and the lower voice follows at a distance of an eighth note.

At the same time, we hear a shift of interval cycle between these two measures. Both upper and lower part state ic2 motion in m. 7, then create ic1 motion in the following measure. The ic1 cycle and the inversional symmetry of m. 8 then continue into the downbeat of m.9, and here the appearance of the two parts is reversed: the LH F# leads, appearing first on the downbeat, postponing momentarily the arrival of its inversional partner D#. With its arrival, this D# then becomes the beginning pitch of the subsequent subject statement in m. 8. This D#, the focal point of the directed motion, becomes set in stark relief, and this moment becomes the dynamic as well as the registral high point of the section.

In addition, the passage holds certain correspondences between contrapuntal and rhythmic structure. The entire fugal passage may be divided into three sections. Measures 1-8 contain the four two-measure entries of the fugal exposition. The second section (mm. 9-11) includes a series of stretti that appear at three- and two-beat intervals. Section 2 is further divided into Sections 2a and 2b. Finally, the third section (mm. 12-13) overlaps the opening fragment of the subject at eighth-note intervals. For visual convenience, the beginning pitch of each entry is circled in the score.

In the exposition, successive statements of the subject and answer appear at the interval of the tritone, but the intervalic pattern shifts in sections 2 and 3. However, Honegger continues to create a consistent and systematic relationship between pitch-class interval and durational interval between successive entries. The following table aligns the two.

Table 3.1: Pitch-class and durational intervals, *Sonatine pour clarinette*

	Exposition	2a	2b	Section 3
Beginning pitch	G C# G C#	D# B F#	D A# E#	G D C# C B A# A
Durational Interval (in eighth notes)	8 8 8	3 3	3 3	1 1 1 1 1 1
Interval class	6 6 6	4 5	4 5	5 1 1 1 1 1

With one exception in Section 3, Honegger maintains a consistent association of durational interval and interval class of successive entries. Within the first section, the entries at ic6 recur at the two-measure level, indicated by the 8 eighth notes in the table. Successive entries in Section 2a and 2b recur at the durational interval of 3 eighth notes, and both 2a and 2b maintain the pattern of interval class 4 and 5. Although the durational interval of 3 eighth notes within both Section 2a and 2b disrupts the notated duple meter, both sections occupy two measures and thereby uphold the higher two-measure metric level established in the exposition. In addition, save for the initial interval class 5 between the first and second entry in Section 3, the subsequent entries pair the interval class 1 with the durational interval of the single eighth note.

The table shows that a characteristic pitch-class interval and durational interval relationship distinguishes each of the three sections, serving to dissociate the sections from one another. On the other hand, a larger two-measure level of organization underlies each of the three sections, creating a metric regularity that binds together the three disparate sections. For the three sections, the two-measure groupings become systematically truncated. The four two-measure units of the exposition diminish to two two-measure units in Section 2, and to a single two-measure unit in Section 3.

Conclusion

Honegger's statement that "everything of use to me is mine"[24] emphasizes the composer's ability to incorporate and work within a multiplicity of idioms and languages. Yet, Kurt von Fischer notes:

> From all his utterances on aesthetic matters and compositional principles, one might think that he would have availed himself of everything, more or less indiscriminately, in realizing this musical thought—that he would have been, then, a distinctly eclectic composer whose style could scarcely be described. Yet such is hardly the case ... Surveying Honegger's work as a whole, one is struck by its compositional-stylistic unity ...[25]

While von Fischer looks to general aesthetic principles to describe this unity, it is also provided, in part, by a sense of musical organization founded in contrapuntal strategies and procedures. This orientation manifests itself both through the use of particular contrapuntal techniques (fugue, imitation, etc.), and contrapuntal symmetries. The latter control the foreground verticalities (through the use of contour, generic interval, and specific interval symmetries) and also organize longer linear spans (through the use of pitch-class inversional symmetry and interval cycles). These strategies and procedures were present from Honegger's earliest compositions, and formed the pedagogical foundation to his studies at the Conservatoire. Moreover, contrapuntal organization in both the vertical and horizontal dimension cuts across the eclecticism of Honegger's pitch and harmonic language, providing one of the common elements in his compositional output.

---

24 Quoted in Marcel Delannoy, *Honegger* (Geneva and Paris: Editions Slatkine, 1986), 85.

25 Kurt von Fischer, "Arthur Honegger," in *Essays in Musicology*, trans. by Carl Skoggard and ed. by Tamara Evans (New York: The Graduate School and University Center, City University of New York, 1989), 159-60.

Chapter 4

Mouvement symphonique (Rugby)

Honegger's second *Mouvement symphonique*, subtitled *Rugby*, was the sole commission of the inaugural concert of the Orchestra symphonique de Paris. The composition premiered under the direction of Ernest Ansermet in 1928. With regard to the subtitle, Honegger explained:

> I very much like football, but I prefer rugby. I find it more spontaneous, more direct and closer to nature than football which is a more scientific game. I am aware of a carefully controlled rhythm in football and for me the savage, brusque, untidy and desperate rhythm of rugby is more attractive. It would be wrong to consider my piece as programme music. All it does is to try to express in my own musician's language, the attacks and ripostes of the game, and the rhythm and color of a match at the Colombes Stadium; I honestly feel it is only right to name my sources. That is the reason why this short composition bears the title of *Rugby*.[1]

Honegger composed three *Mouvements symphoniques*. The genre finds its origin in the nineteenth-century symphonic poem, a single-movement orchestral composition based upon a nonmusical idea, often literary or philosophical. In France, more recent antecedents included Debussy's *Prélude à l'après midi d'un faune* and Dukas' *L'Apprenti sorcier*. Prior to writing his three *Mouvements symphoniques*, Honegger composed two symphonic poems while a student at the Conservatoire. The first, the *Prélude pour Aglavaine et Sélysette* (1917), was a Debussyist composition based upon a literary work by Maurice Maeterlinck. The second, influenced by Strauss, was *Le Chant de Nigamon* (1917), based upon an episode from Gustave Aimard's novel *Le Souriquet*, and which incorporated several Native American themes taken from Julien Tiersot's *Notes d'ethnographie musicale*. Yet, Honegger's later *Mouvements symphoniques* suggest a renovation of the symphonic poem: the first two (*Pacific 231* and *Rugby*), "realistic glorifications of the modern age, represent, both in subject and in their percussive and motoric idiom, an attempt to bring new life to

---

[1] Translated and quoted in Geoffrey Spratt, *The Music of Arthur Honegger* (Cork: Cork University Press, 1987), 149.

the symphonic poem ..."[2]

As in the case of the first *Mouvement symphonique (Pacific 231)*, Honegger downplayed the programmatic aspects of the composition, partially in reaction to critics who considered both works onomatopoeic.[3] As a result of this reaction, perhaps, the third *Mouvement symphonique*, completed five years after *Rugby* in 1933, bore no subtitle at all.

These three one-movement orchestral works, written between 1923-1933, represent an early stage of Honegger's compositional career. The composer turned only later to three-movement symphonies. Although his first symphony was completed in 1930, his last four symphonies were written at a later stage, between 1941 and 1950, well after the three *Mouvements symphoniques*. This very gradual shift from one-movement to three-movement orchestral works probably reflects an early difficulty with absolute symphonic works which, he stated, "demand an effort at sustained reflection."[4] On the other hand, Honegger claimed compositional facility when inspired with a visual or literary idea.

An important stylistic feature of Honegger's three *Mouvement symphoniques* is the emphasis upon rhythm, and the works feature metric shifts and emphasize rhythmic conflict more than do the symphonies. As a rhythmic essay, *Rugby* resembles *Pacific 231*, although the absence of percussion in this second *Mouvement symphonique* distinguishes it from the first.[5] The composer contrasted the calculated acceleration process of *Pacific 231* to the more malleable rhythmic language of *Rugby*:

> I had written a first *Mouvement symphonique* in which I tried to produce an increase in the sense of speed with a decrease of tempo—a progressive gradation of rhythmic structure in the form of a great chorale developed by J. S. Bach. As a subject, I had taken the locomotive of the *Pacific* type, indicated as 231 for heavy rapid-speed trains.
>
> In *Rugby, Mouvement symphonique No. 2*, I was taken by an idea exactly the opposite. To the quasi-mathematical progression of the machine, I wished to oppose the diversity of human movement: its brusque surges, its stops, its flights, its flaggings.[6]

---

2 Willi Apel, ed. *Harvard Dictionary of Music*, 2d ed. (Cambridge, Mass.: Harvard University Press, 1977), s.v. "Symphonic Poem."

3 See, for example, Honegger's jeremiad in *I am a Composer*, trans. Wilson O. Clough (New York: St. Martin's Press, 1966), 100-101.

4 Honegger, *Composer*, 81.

5 Also curious in *Rugby* is the complete absence of dynamic markings following the *f sempre* in m. 2.

6 "J'avais écrit un premier *Mouvement symphonique* dans lequel j'essayais de rendre l'accroissement du sentiment de vitesse par un ralentissement du temps. Une gradation progressive de la structure rhythmique dans la forme du grand choral développé par J.S. Bach. Comme sujet, j'avais pris la locomotive type *Pacific*, indice 2-3-1, pour trains lourds de grande vitesse.

The "surges, stops, flights, and flaggings" are in part represented by the procedure of juxtaposition and interruption of short compositional ideas. This technique recalls Stravinsky, but the contrapuntal and formal processes follow a clearer and more traditional trajectory. The form of the composition is a rondo with two main themes.[7] The large-scale formal outline is summarized in Figure 4.1.

Figure 4.1: Formal outline of *Mouvement symphonique, no. 2*

M. 1	Introduction
12	Theme 1
27	Episode
85	Theme 1 (extended)
122	Introduction
127	Theme 2
144	Episode 2
242	Theme 1
250	" "
261	Coda (Intro and Theme 1)

The rondo sections (Themes 1 and 2) are given over to diatonicism with arpeggiated triadic themes in the brass. These sections alternate with episodes of more rhythmic complexity. In discussing the inspiration for the piece, Honegger identified three constructive elements: theme, rhythm, and contour. Citing three visual points of departure, he noted:

> To each of these visual or perceptible elements corresponds a theme, a rhythmic motive, or a melodic contour. Between them, they play according to the laws of a musical rugby, and attempt to build the piece following the rules of a symphonic game whose development and progression maintain the same rigor as a piece of music deemed 'absolute.'[8]

"Dans *Rugby, Mouvement symphonique No. 2*, j'ai été poussé par une idée exactement contraire. A la progression quasi-mathématique de la machine, j'ai voulu opposer la diversité du mouvment humain: ses brusques élans, ses arrêts, ses envolées, ses fléchissements."

Arthur Honegger, *Ecrits*, Textes réunis et annotés par Hughette Calmel (Paris: Librairie Honoré Champion, 1992), 170. Originally appeared in an essay entitiled "La musique et le sport" from the *Revue officielle du Commissariat Général*, juillet-août-septembre 1943.

7 The analytical literature for the composition is scant; for two short analyses see Harry Halbreich, *Arthur Honegger: Un musicien dans la cité des hommes* (Librairie Arthème Fayard, 1992), 433-5; and Spratt, *Arthur Honegger*, 148-61.

8 "A chacun de ces éléments visuels ou sensibles correspond un thème, un motif rhythmique ou un contour mélodique. Ceux-ci jouent entre eux selon les lois d'un *Rugby musical*, et s'efforcent de construire le morceau d'après les règles du jeu symphonique et avec la même

Example 4.1: Mm. 1-7

Together with theme and rhythm, then, contour supplies a fundamental constructive element to the work. This element often structures and works in tandem with the rhythmic gestures. A characteristic strategy involves the use of clear unidirectional contours that complement and dilute the complexity of rhythm. Example 4.1 contains the opening measures of the composition, and shows how rhythm and contour interact. These opening seven measures form three gestures, distinguished by contour and rhythmic character. These are labeled X, Y, and Z in the score. X, the first gesture, occupies the first two measures, forming an ascent that connects the beginning C#4 to D6 on the downbeat of m. 3. Y corresponds to the following eighth-note triplet gesture.

---

rigueur dans le développement et la progression que lorsqu'il s'agit d'une oeuvre de musique dite 'pure.' " Honegger, *Ecrits*, 170.

This motive reverses direction, creating an overall descent until the last beat of m. 5. Dotted slurs in the upper part show how Y is comprised of a six-note motive stated four times, each time an octave lower. Z begins at the last beat of m. 5 and here the shift in direction corresponds to a change in rhythmic subdivision to sixteenth notes. Again, dotted slurs show that Z consists of a six-note motive stated four times, each time an octave higher. Note that the contour reversal that takes place between Y and Z underlines the alternations in rhythmic subdivisions.[9] Thus, in these opening measures, contour and rhythm participate together in structuring the three gestures of X, Y, and Z. The shifts in rhythmic subdivision correspond to shifts in contour.

Introduction (Mm. 1-7)

The composition is indebted to the procedures characteristic of the symphonic poem, especially that of thematic transformation, which provides a means for musical development with consistent motivic material. The analyses will show that, in *Rugby*, the notion of thematic transformation becomes expanded to include *abstract* pitch and rhythmic relationships. These abstract relationships create further levels of cross-reference in the composition.

The analyses will also show that pitch and rhythmic elements stated within the introduction give rise to a significant amount of musical material within the body of the composition. In addition to the overt thematic repetitions and transformations that take place (such as the return of both Rondo themes), Honegger develops more subtle resonances out of these opening measures. Many of these correspondences arise out of pitch material, rhythmic groupings, and rhythmic and metric conflicts stated in the first seven measures. These are then expressed and expanded later in the work. Moreover, particular rhythmic conflicts within the introduction are not only transferred to higher levels of rhythmic structure as the composition progresses, but also become stated in the domain of pitch.

Motive X (Mm. 1-3): Pitch, Rhythmic Grouping

Example 4.1 shows that the opening three measures link C# to D, a leading-tone gesture which moves to resolution. The pitch focus expresses the overall harmonic evolution of the composition, which ends unambiguously in D major.

[9] Y and Z are not exclusively unidirectional, however, and Honegger uses contour to set in relief the boundaries of the repeated six-note motives. For example, within the overall descent of Y, the whole-step ascent of C-D delineates the boundaries of the six-note motive; within the ascent of Z, a half-step descent between D-C# again sets off the boundaries of the motive. Additionally, Z contains a G-E descent between the second and third pitches of the motive.

The first two measures form a structural upbeat to m. 3.

In addition, the trumpets rearticulate the opening C# on each beat of the first measure, allowing a perception of the quarter-note pulse before yielding to the ascent in m. 2. The first two measures then parse into two sub-gestures: the repeated C# is followed by the scalar ascent, forming a rhythmic grouping of 4 + 2 quarter notes. This grouping contradicts the notated 3/4 meter and will form a significant abstract rhythmic motive throughout the first half of the composition.

Motives Y and Z (Mm. 3-7): Rhythmic Conflicts, Contour

The relationship between Y and Z, the second and third gesture, creates conflict on both the level of rhythmic subdivision and meter. Most evident is the 3 : 4 conflict of rhythmic subdivision between Y and Z, moving from eighth-note triplets to sixteenth notes. As indicated above, this shift from eighth-note triplets at Y to sixteenth notes at Z is underlined by the change in contour.

In addition to this conflict of rhythmic subdivision, the motives in both Y and Z (indicated by dotted slurs in the score) suggest an implied metric conflict. The repeated six-note groupings of Y and Z express a meter at odds both with the notated meter and with each other. The repeated motive at Y suggests a duple meter of 2/4, supplanted by an implied level of 3/8 at Z.[10]

Within the larger descent of Y, the pairs of accompanimental chords assert an inversionally symmetrical relationship. As the dotted slurs in the lower part of Example 4.1 indicate, the upper voice of the dyad ascends by ic3, and the lower voice descends by ic3. Here, this local symmetry is subordinate to the larger overall descent of Y. This provides a strategy of local inversional symmetry within an overall context of a larger-level unidirectional contour. This strategy is one that recurs throughout the composition.

The events within the introduction are summarized below in Figure 4.2.

Figure 4.2: Mm. 1-7 of *Rugby*

1) Leading-tone motion of X, linking C# to D
2) 4 + 2 quarter note rhythmic grouping of X
3) 3 : 4 (eighth note triplet vs. sixteenth note) conflict of rhythmic subdivisions between Y and Z
4) 2/4 : 3/8 metric conflict between Y and Z
5) Contour reversals between Y and Z underline the conflict of rhythmic subdivision; unidirectional contours despite the use of local inversional symmetries

---

[10] The implied meter at Y is duple. A metric reading of 6/8 is also possible at Y.

The analysis examines the recurrences of these elements within the first half of the composition before proceeding to the second half, where the opening rhythmic conflicts emerge onto a higher level of structure. Figure 4.3 contains the principal sections of the first half of the composition.

Figure 4.3: Principal sections of the first half of *Rugby*

Rehearsal	Measure	Section
	M. 1	Introduction
1	M. 12	Theme 1
2	M. 27	Episode 1: Section 1
3	M. 33	Section 2
4	M. 43	Section 3 (Derived from Section 1)
5	M. 57	Section 4
6	M. 67	Section 5
7	M. 74	Section 6 (Derived from Section 4)
8	M. 85	Theme 1
12	M. 122	Introduction
12 + 5 mm.	M. 127	Theme 2

Metric and Pitch Recurrences: Mm. 111 ff.

To begin, we examine a section that contains a varied return of mm. 3-7 from the introduction. This varied return appears at m. 122 (R. 12). Example 4.2 begins at R. 11 and includes this section. Motive Y appears at m. 122, at T7 from its statement in the introduction. Z follows at m. 125, retaining its rhythmic identity of sixteenth notes and its overall ascending contour, but losing its pitch relationship to mm. 5-7.

The material between Rehearsal 11 and 12—even prior to this m. 122 recurrence of Y and Z—recalls some of the opening material. A metric conflict appears that restates the metric conflict between Y and Z from mm. 3-7. Beginning in m. 112, both the bass and the upper parts assert an implied duple level of 2/4. This metric layer continues throughout the section. At m. 118 the metric conflict intensifies, and stated against the 2/4 of the outer parts occurs a layer of 3/8, articulated by the [027] chords, given to the trumpets and oboes in the orchestral score. These two metric levels recapitulate the 2/4 and 3/8 meters earlier implied by Y and Z at mm. 3-7. In contrast to mm. 3-7, however, these levels of 2/4 and 3/8 are now stated simultaneously rather than successively.

In addition, the pitch organization of the passage leading to Rehearsal 12 recalls and enhances the leading-tone motion of mm. 1-3. At mm. 118-121,

Example 4.2: M. 111 ff. (R. 11-12)

Example 4.2 (Continued)

Example 4.2 (Continued)

the upper part rearticulates Ab6 before ascending to A at m. 122. At the same time, the bass states Eb2 and Eb3 before descending to D. Within mm. 118-122, then, Honegger transfers the half-step motion to both outer voices, structuring the m. 122 arrival through inversionally symmetrical motion.

In a larger context, the harmonic motion likewise alludes to the pitch organization of mm. 1-3 of the introduction. The Db-based diatonic tonality of Theme 1 at m. 111 cedes to the arrival upon D at the return of the introduction (m. 122). This harmonic progression magnifies the initial C#-D leading tone gesture of X which opens the composition. Additionally, the arrival points at m. 111 and m. 122 are themselves framed by an earlier statement of Theme 1 (m. 85) in Db major and a later statement of Theme 2 in D major (m. 127). The large-scale tonal direction of these sections which closes the first half of the composition, despite an interruption between mm. 93-110, projects the C#-D melodic motion of the introduction. This harmonic motion is summarized in Figure 4.4.

Figure 4.4: Harmonic motion of mm. 85-144

Theme 1	Mm. 85-92	Db Major (diatonic)
Interruption	Mm. 93-110	
Theme 1	Mm. 111-121	Db Major (diatonic)
Introduction	Mm. 122-127	D (arrival in bass)
Theme 2	Mm. 128-143	D Major (diatonic)

Episode 2	M. 144	

In mm. 85-144, which closes the first half of the composition, several elements from mm. 1-7 return. The restatement of the introduction (m. 122) involves the return of Y (eighth note triplets) and of Z (sixteenth notes). Moreover, the 2/4 : 3/8 metric conflict between Y and Z is played out at mm. 112-122. Finally, the successive key areas of Db and D at mm. 85-143 magnify the C#-D leading tone motion of X, initially stated in mm. 1-3.

4 + 2 Rhythmic Groupings: Theme 1 and Episodes

Thus, the leading tone gesture of Motive X (mm. 1-2) relates to the later succession of keys at mm. 85-143. In addition to resonances with the pitch organization of X, the 4 + 2 rhythmic grouping of this mm. 1-2 motive similarly reappears within the composition. Despite the presence of contrasting pitch and motivic material, this 4 + 2 rhythmic grouping functions as an abstract rhythmic motive, and initiates *each* section and subsection of the first half of the composition.

Example 4.3: Theme 1, mm. 12-17 (R. 1)

Example 4.3 contains the first appearance of Theme 1 at m. 12 (R. 1). This statement of Theme 1 serves to clarify the metric ambiguity of the first eleven measures of the composition and to integrate the music with the notated 3/4 meter. Yet this metric clarification is withheld for the first two measures. In the bass, durational accent accrues to the downbeat of m. 12 and to the second beat of m. 13, connecting G, through the chromatic neighbors G# and Bb, to A. The melodic material likewise withholds the downbeat accent at m. 13, and dramatizes the approach of the second beat through the anticipation of a sixteenth-note. Here the rhythmic grouping of mm. 12-13 echoes that of the

opening two measures of the composition, partitioning six beats into a grouping of 4 + 2 quarter notes.

Similarly, the beginning of the first episode at m. 27 (R. 2) alludes to this 4 + 2 rhythmic motive. Example 4.4 contains the opening six measures of the first episode.

Example 4.4: Episode 1, mm. 27-32 (R. 2)

Note that the initial three-measure idea (mm. 27-29) returns at T1 in the following three measures (mm. 30-32). The dividing point of both of these three-measure groupings occurs on an agogic, contour, textural, and notated accent which overshoots the downbeat of their second measure by a beat. Brackets above the score indicate this. In the melody, this accentual division of each three-measure phrase partitions the constituent rhythmic subdivisions, separating the sixteenth notes from the eighth-note triplets. Moreover, the divisional accents within both three-measure groups coincide with the intervalic mid-point of the ascending ninths in the bass, which bisects the span from A1 to B2 with E2 (mm. 27-29), then the span from Bb1 to B#2 with F2 (mm. 30-32). This dividing point of the three-measure motives echo the 4 + 2 grouping of X and Theme 1. Now, however, Honegger extends the second portion of each gesture, creating three measure groupings. This results twice in a grouping of 4 + 5 for these six measures.[11]

The motivic material for Section 4 displays the same process of rhythmic acceleration employed in Honegger's *Pacific 231*, compressed here into ten measures. In fact, the sketches for *Pacific 231*[12] contain an embryonic form of this motive, but it was not used in that earlier composition itself. Example 4.5 displays the motive from the sketches, as well as its appearance in the score of *Rugby*, beginning at m. 57 (R. 5). The process of rhythmic transformation upon a six-note cell in the sketch recalls Y and Z of the introduction, where six-note motives appear first within eighth-note triplets, then sixteenth notes.

Honegger maintains the general process of rhythmic crescendo from the sketches, but he significantly develops and extends this motive in the published score. In the sketch, the six-note motive is repeated twice at each level of rhythmic subdivision. In the score, however, this motive appears in m. 58, embedded in the second measure of a four-measure group that begins in m. 57 (the arrow in the example shows its appearance in the score). Within this four-measure group of mm. 57-60, the melodic motion of the first and third measures is identical (m. 57 and m. 59), while the second and fourth measures (m. 58 and m. 60) are kindred but not identical. In relation to the sketch, the motive, and therefore the entire passage in the score, now doubles in length, from five to ten measures.

A general contour inversion arises between the first two measures, where the opening three pitches at m. 57 form a Contour Class (CC) of < 0 1 2 >,

---

[11] The six-measure organization of this section shown in Example 4.4 responds to the six-measure organization of the preceding statement of Theme 1 (mm. 12-17). Their organization differs, however. On the one hand, the repeated motives at mm. 27-32 twice establish a three-measure hypermeter. On the other hand, at mm. 12-17 the returning G in the melody asserts a two-measure hypermeter. (See Example 4.3.) Other hypermetric conflicts are discussed further in the chapter.

[12] Contained at the Paul Sacher Stiftung, Microfilm 186: 788-789.

and at m. 58 form < 2 1 0 >.[13] The disturbance of the exact contour inversion occurs on the second beat of m. 58, coinciding with a return to the initial pitch of

Example 4.5: Compositional sketches and score, m. 57 ff. (Section 4)

[13] A contour class assigns numerical values to a succession of pitches, with the lowest pitch set at 0. For analytical presentations of musical contour, see Michael Friedmann, "A Methodology for the Discussion of Contour: Its Application to Schoenberg's Music," *Journal of Music Theory* 29 (1985): 223-48; Elizabeth Marvin and Paul Laprade, "Relating Musical Contours: Extensions of a Theory for Contour," *Journal of Music Theory* 31 (1987): 225-67; and Robert Morris, "New Directions in the Theory and Analysis of Music Contour," *Music Theory Spectrum* 15/2 (Fall 1993): 205-28.

Example 4.5 (Continued)

Copyright 1928 by Editions Senart; piano version pub. 1929. Used by permission of Editions Salabert.

Bb, written as A# in the score. Beginning with this A#, the CC subsequently repeats m. 57's opening ascent of < 0 1 2 2 >. Example 4.6 shows the repeated contour segment that appears both in m. 57 and m. 58: the return of the CC and of the pitch Bb/A# provides a sense of "beginning again."

The return to A# at m. 58 partitions the notated six beats into a grouping of 4 + 2. As in the previous sections, the opening of this section again restates the 4 + 2 abstract rhythmic motive of X. Significantly, Honegger tailors the motive from his sketches, and this alteration accords with the 4 + 2 rhythmic motive.

Additionally, a fourth pitch, C, is added to the three discrete pitches from the sketch (D-C#-A#) in mm. 57-58. This added C3 contributes to the structuring of voice relationships between both parts at mm. 57-60. Example 4.7 shows that the unfolding of pitches in the melody—Bb, C, Db, and, in the following measure, D—mirrors the descent in the bass. Progressing at the level of the measure, the bass moves from Bb, Ab, G, to Gb. While the inversional relationship between voices is rhythmically displaced, the relationship of the voices creates pitch inversional symmetry, matching the interval ascent of < +2 +1 +1 > in the upper voice with the interval descent of < -2 -1 -1 > in the lower.

The continuation of Example 4.7 shows that Honegger maintains inversional symmetry in the following measures, beginning at m. 61. With the continuation of the motive after the initial four measures, the lower part is stripped of stepwise motion, preserving its boundary of -4 intervals. The bass now rhythmically coincides with its inversional partner in the upper part, which states +4. Here Honegger preserves inversional symmetry, matching the C#-F (mm. 61-62) of the upper part with Db-A, F-A with F-C# (mm. 63-64), and finally G#-C with Ab-E (mm. 65-66).

Finally, the transpositional levels effected onto the motive at the point of rhythmic transformation likewise bring about symmetry in pitch space. While the motive is stated four times, beginning with Bb, C#, F, and G#, the shift of rhythm occurs between the first two and the last two statements, straddling the Bb-C# and the F-G# motives. Example 4.8 displays the traversal of fifteen semitones in pitch space at these junctures, which grafts a higher-level unidirectional contour onto the localized inversional symmetries. Contour thus structures the overall direction of the ten-measure passage.[14]

The previous examples suggest that many of the salient pitch and rhythmic events stated within the mm. 1-7 introduction provide a basis for the construction of subsequent events within the first half of the composition. Associations are developed out of rhythmic groupings, rhythmic/metric conflicts, and pitch materials, and these recurrences take on a motivic function. We have seen that, within the dimension of pitch, the leading-tone motion not only heralds the return of the introduction at m. 122, but determines the harmonic structure of mm. 85-143, in which a Db tonality is supplanted by the move to D major with the arrival of Theme 2 at m. 128.

---

14 Moreover, the superimposition of an overall single-direction contour upon local inversional symmetries, as shown by Examples 4.7 and 4.8, is a strategy observed within Motive Y of the introduction.

Example 4.6: 4 + 2 grouping, mm. 57-58

Example 4.7: Inversional symmetry, m. 57 ff.

Example 4.8: Larger-level ascent of Section 4

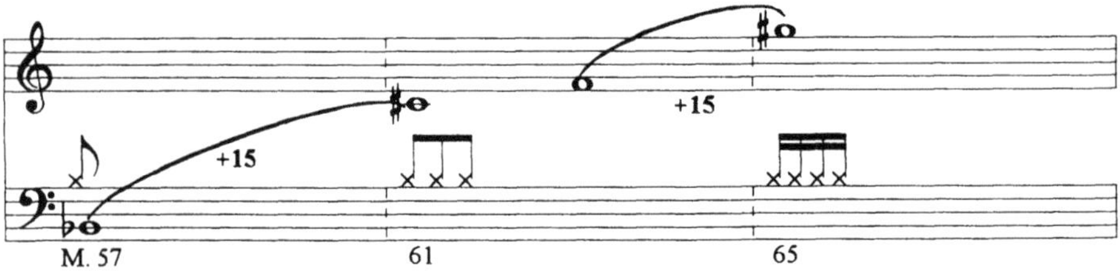

Additionally, the returning rhythmic motive, grouping 4 + 2 quarter notes, returns and begins each section of the first half of the composition. Within the second half of the composition, other events of the introduction, especially the 3 : 4 conflict of the rhythmic subdivisions between Y and Z, create subsequent material.

Metric Organization: M. 144 ff.

The second half of the composition is defined by the shift to 4/4 meter, and bisects the composition nearly exactly, occurring at m. 144 of its 291 measures. The alternation of 4/4 and 3/4 meters begins properly with the double bar at m.

186 (R. 19), and continues until the end of the composition.[15] Figure 4.5 displays these metric shifts. The first three pairs of metric shifts define a proportion of 2 : 1 for the number of measures (8 : 4; 6 : 3; 4 : 2), as well as a systematic decrease in length. Subsequent sections are less systematically proportioned.

Figure 4.5: Metric shifts in the second half of *Rugby*

Measure	Number of mm.	Meter
186	8	4/4
194	4	3/4
198	6	4/4
204	3	3/4
207	4	4/4
211	2	3/4
213	2	4/4
215	3	3/4
218	8	4/4
224	1	3/4
225	2	4/4
227	1	3/4
228	2	4/4
230	1	3/4
231	7	4/4
238	2	3/4
240	21	4/4
261	18	3/4
279	5	4/4
284	7	3/4

Correspondences of Meter and Transpositional Cycles

Since the first half of the composition remains in 3/4, metric organization in the second half departs significantly from that of the first half. In one sense, perhaps, the metric ambiguity supplied by the 4 + 2 rhythmic motive in the first half of the composition signals the shift to quadruple meter which follows. In another sense, however, the conflict of rhythmic subdivisions at Y and Z—the conflict between triple and quadruple divisions of the beat—projects onto another layer of rhythmic structure. Thus, these levels of rhythmic subdivision become stated on

[15] The double bar is given in Honegger's piano transcription, but not in the orchestral score.

the metric level.[16] The transfer of this 3 : 4 conflict to a higher level of structure then governs the unfolding of the second half of the composition, which shifts between 3/4 and 4/4 meter.

Additionally, this conflict becomes expressed in the dimension of pitch. Example 4.9a contains the 3/4 sections between mm. 194-212. These comprise the second, fourth, and sixth metric shifts of the second half of the composition, in which the sections become successively truncated. Each of these 3/4 sections contains a recurrence of the melody from Th. 1. Accompanying this melody appears an ascending triadic figure progressing in quarter notes and, below, an eighth-note motive in the bass. In each measure the bass motive states a chromatic wedge gesture that is inversionally symmetrical around its first two pitches. This motive is transposed downward at the downbeat of each measure, providing an overall descending contour that overrides the local inversional symmetry of the wedge gesture within each measure. These two accompanimental parts within the 3/4 passages, the bass and the midrange triads, clearly establish a conflict of contour, since the ascent of the triads is mirrored by the descent in the bass. In this passage, the conflicting contours are asserted differently: consecutive statements of the triads ascend on each beat, while the bass descent appears at the level of the measure.

Example 4.9b displays the downbeat pitches of both accompanimental gestures. This measure-level descent in the bass appears primarily at the directed interval of <-3> in each occurrence. Within the first 3/4 statement at mm. 194-198, the directed interval series of <-3 -3 -3 (-3)>[17] connects G#2 to Ab1, the second statement (mm. 204-207) contains two pairs of ic3 descents, stating <-3 -1 (-3)>, and the final appearance of the bass motive between mm. 211-212 likewise descends at <-3>, from C#3-A#2. At measure downbeats, the accompanying ascending triadic gesture is stated primarily at the directed intervals of +4: the <+4 +6 +6> at mm. 194-198 becomes replaced by <+4 +4> within mm. 204-207, and <+4> between mm. 211-212.

The conflict of directed interval cycles mirrors the 3/4 : 4/4 metric conflict in the second half of the composition. In the successively truncated 3/4 sections, the measure downbeats descend by three semitones in the bass and ascend four semitones in the accompanimental upper part, projecting into directed pitch space the 3 : 4 conflict.[18]

Example 4.10a reveals a similar strategy at mm. 235-241. The motive in this passage is derived from m. 27 (shown in Example 4.4), which

---

[16] The notion of relationships developed across different levels of rhythmic structure is discussed extensively in Ch. 2.

[17] Parentheses indicate the downbeat of the following 4/4 section.

[18] There is also evidence of a third interval cycle within the melodic statement at mm. 204-207, where the downbeat pitches progress from F4-C5-G5-D6. This ic5 cycle is independent of the metric correspondences discussed here.

Example 4.9a: 3/4 sections, mm. 194-212

Example 4.9b: Metric reduction of 3/4 sections

originally composed out a 4 + 5 metric grouping for three measures of 3/4. Beginning in mm. 235 (R. 24), however, the motive is shortened in its second half, and the resultant 4 + 4 grouping corresponds to the notated meter in this 4/4 section. At each measure the motive is stated in counterpoint with itself. As in its original statement, the two halves of the motive assert conflicting rhythmic subdivisions, and here within the two-measure motive, the first measure states sixteenth notes, succeeded by eighth-note triplets. As a result, the statements that occur each measure in stretto now verticalize this sixteenth/eighth-note triplet conflict.

Within the first five measures, each statement descends at a distance of three semitones per measure.[19] Example 4.10b contains the beginning pitches of each statement, in both the bass and the upper voices. Five statements of the T-3 cycle occur. This cycle is not only completed but begun again. M. 239 returns to the same pitch classes as m. 235. The subsequent change to 3/4 meter at m. 240 coincides with a change in transpositional cycle. Three statements within these two measures appear two beats apart, descending at T-4.

[19] The transpositional cycle is in pitch class, not pitch space: the beginning pitch of each statement includes C4, A3, F#3, Eb4, C4.

Example 4.10a: Mm. 235-241

Example 4.10a (Continued)

Example 4.10b: Reduction of mm. 235-241

The transpositional levels *reverse* the metric levels during the passage. Between mm. 235-240, the 4/4 meter harbors the T-3 cycle; the following 3/4 meter contains the T-4 cycle.[20] Again, Honegger activates the 3 : 4 conflict—initially expressed on the level of rhythmic subdivision in the introduction, and stated on the metric level in the second half of the composition—through the transpositional cycles in the pitch dimension.

Return of Theme 1 and Introduction

The rhythmic conflicts of the composition culminate with the return of Theme 1, which appears twice consecutively before the return of the introduction. Theme 1 is stated first at m. 242, and repeated at m. 250.[21] Example 4.11 contains the m. 250 restatement of Theme 1, which effects a polymetric conflict in this section. Note that the repeated motive in the upper part clearly asserts the notated 4/4 meter, with the accompanimental motive (upper staff) recurring rhythmically identically in each measure.

Against this in the bass, however, an implied metric level of 3/4 gradually emerges. Initially this metric identity is ambiguous, and the metric groupings in the bass are partitioned as 3 + 2 + 3 + 2 + 2 between mm. 250-252. Beginning in m. 253, though, the bass unambiguously states triple groupings, maintained until m. 260, two beats before the return of the introduction at R. 27. Example 4.11 brackets these groupings.

Given this implied 3/4 metric level in the bass against the 4/4 meter of the upper accompanimental part, the second statement of Theme 1 between mm. 250-260 now states simultaneously the 3 : 4 metric conflict that takes place successively throughout the second half of the composition.

In addition, between the two statements of Theme 1 at m. 242 and m. 250a parallel hypermetric conflict also obtains. Example 4.12 indicates this. The first section at m. 242 contains four-measure groupings, determined by the melody phrases initiated in parallel fashion. Both m. 242 and m. 246 begin with the pitch class D on their second beat, thus creating two four-measure groupings for the eight-measure section. In contrast, the following statement of Theme 1 at m. 250 occupies three-measure groups for the first nine measures of this eleven-measure section. The hypermetric groupings at mm. 250 ff. are made clear by the return to D in the bass each three measures. (The triple groupings are additionally reinforced by the repeated three-measure accompanimental gesture in

[20] Arguably, the 3/4 notated meter does not here represent the expressed meter of the passage, which may perhaps be better represented by 3 measures of 2/4. Nevertheless, the correspondences between the notated meter and transpositional levels is clear.

[21] For a discussion of several contour symmetries operating within the m. 242 statement of Theme 1, see Chapter 3.

the upper voice. Compare the upper stave of mm. 253-255 with that of mm. 256-258 in Example 4.11.)

Example 4.11: Restatement of Theme 1, Mm. 250-261

Example 4.11 (Continued)

Example 4.11 (Continued)

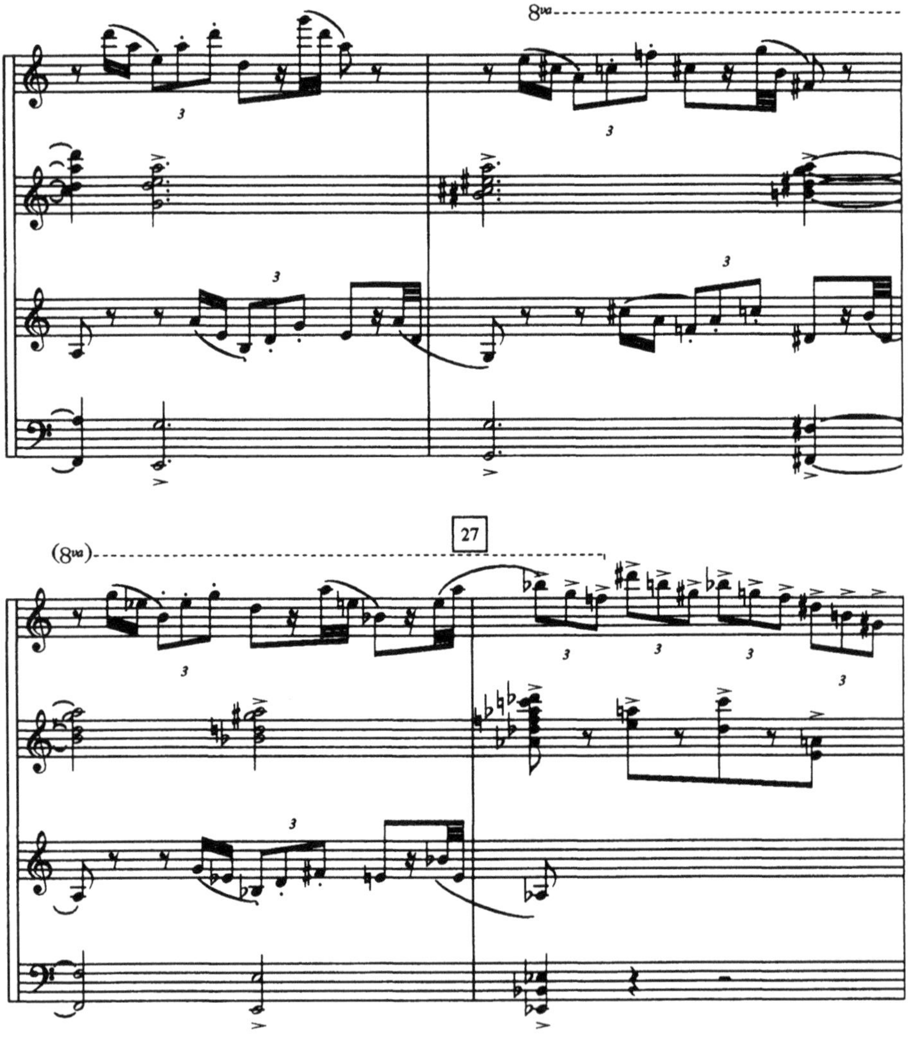

Example 4.12: Hypermetric conflict at Theme 1,
mm. 242-249 and mm. 250-260

Hence, while both statements at R. 25 and 26 allude to Theme 1, their hypermetric organization varies. As Example 4.12 shows, mm. 242-249 (R. 25) is stated as 4 + 4, and mm. 250-260 (R. 26) as 3 + 3 + 3 + 2. This promotes to the hypermetric level the 3 : 4 metric conflict that organizes the second half of the composition. The hypermetric conflict between these two successive statements of Theme 1 allows the composer to reconfigure repeated thematic material, and to create continuity with this thematic material while at the same time creating development and conflict through the metric material. In addition, the 3 : 4 hypermetric relationship between these two statements of Th. 1 extends to yet a higher level of metric structure the primary conflict of the second half of the composition.

The D major tonality of the two statements of Theme 1 (mm. 242 and 250) is supplanted by the move to Eb in the bass at m. 261. (This occurs in the last measure of Example 4.11.) The motion restates at T1 the tonal motion from Db to D in the parallel section between m. 111-122,[22] and a comparison of these two parallel sections that prepare this half-step motion reveals interesting correspondences. Example 4.13 contains the initial measures of both sections, aligning mm. 111-118 of the first half of the composition with mm. 250-255 of the second half.

Both sections begin with nearly identical stepwise motion in the bass, as well as with the identical melodic material, which rearticulates the root and fifth of the respective keys. In both sections, this material does not conform to the notated meter.

---

[22] See earlier in this chapter.

Example 4.13: Mm. 111-118 and mm. 250-255

Moreover, the notated meter differs within each section. Therefore, at one measure past m. 111, an expressed duple meter is asserted against the notated triple meter. Conversely, in the later section, an expressed triple meter operates against the notated duple, beginning at m. 253. (As noted above, the triple groupings begun at m. 250 do not become consistent until m. 253.) This inconsistency of grouping between mm. 250-252 allows for a three-measure grouping that parallels that of the earlier section (shown directly above in Example 4.13). The Db initiated in the bass at m. 111 returns at m. 114. Similarly, the D initiated in the bass at m. 250 returns at m. 253.

The parallel organization of both sections seems to cease in the measures following. Example 4.14, which now aligns mm. 118-122 with 256-261, shows that between mm. 118-121 the lower part repeats the same harmonies against the [027] chords. On the other hand, between mm. 256-260, the stepwise bass motion and motivic counterpoint are retained from the previous measures. The half-step motion in the outer voices is very clearly articulated between mm. 118-122, shown by the dotted slurs. Eb moves to D in the bass, and Ab moves to A in the upper part. However, this motion is merely relegated to the final measures at mm. 260-261, shown by the dotted slurs. E moves to Eb in the bass, while A moves to Bb in the upper part.

Yet despite the difference in pitch and motivic organization, an abstract rhythmic relationship takes place. The two simultaneous metric levels which begin at m. 118 (2/4 and 3/8) establish a 3 : 4 conflict at the level of the eighth note. At m. 256, the two simultaneous levels of metric

Example 4.14: Mm. 118-122 and mm. 256-261

organization, 4/4 and 3/4, likewise draw a 3 : 4 conflict. Now, however, this conflict appears at the level of the quarter note.

Furthermore, the relationship between these two sections is one of *exact* augmentation. Example 4.14 reveals that, in both sections, one level articulates a grouping of 4 + 4 + 4 + 4 + 4; against that level is stated the grouping of 3 + 3 + 3 + 3 + 3 + 3 + 2. Interestingly, the role of these groupings is reversed. In the earlier section the bass articulates the grouping of four, while an accompanimental upper voice states that of three. In the later section the bass grouping is triple, while the upper accompanimental level is quadruple.

These two sections fulfill the same formal role of bringing about a return of Y and Z in the composition, first at m. 122 and, later, at m. 261. Although the motivic relationship between mm. 111-118 and mm. 250-255 is clear (see Example 4.13), the motivic association between sections is lost in the following measures. Instead, as Example 4.14 has shown, rather than a motivic relationship, Honegger instead advances an abstract rhythmic correspondence between mm. 118-122 and mm. 256-261.

Conclusion

The composition is indebted to the procedures of thematic transformation characteristic of the symphonic poem. Overtly, thematic material often recurs altered, elaborated, and juxtaposed. These recurring themes are often cast with different metric identities, so transformation is also extended into the metric domain.

In addition to these patent thematic recurrences and transformations, Honegger provides musical correspondences through other relationships. The analyses show that Honegger often bases motives upon abstract rhythmic groupings and conflicts. If less overt, these materials weave subtle musical relationships into the fabric of the composition. Material derived from the opening Motives X, Y, and Z is used throughout the composition. Motive X contributes the C#-D leading-tone gesture, which reflects the large-scale succession of keys at mm. 85-144. This motive also contains the 4 + 2 rhythmic grouping that opens the subsections of the first half of the composition (at mm. 12, 27, and 57).

The eighth-note triplet grouping of Motive Y and the sixteenth-note grouping of Motive Z generate the conflict of 2/4 and 3/8, and these two meters return, juxtaposed, at mm. 118-121. In addition, the 3 : 4 conflict of these eighth-note triplet/sixteenth-note Motives Y and Z relates to the 3/4 : 4/4 metric conflict in the second half of the composition. The transpositional cycles used in the second half (mm. 194-212 and 235-241) reflect this 3 : 4 conflict. Moreover, the second half also exploits the 3 : 4 conflict in the realm of hypermeter. The return of Theme 1 at mm. 242-249 uses four-measure groupings, while the return of the same theme directly following (mm. 250-260) uses three-measure groupings.

Chapter 5

Symphonie pour cordes

This chapter examines the first and second movements of Honegger's *Symphonie pour cordes*, considering both the published score and Honegger's compositional sketches for these two movements. The contents and the ordering of the compositional sketches illustrate Honegger's compositional methods and reveal some correspondences worked into the piece, particularly in the second movement. The earliest correspondences are intervalic, before Honegger moves on to develop pitch and pitch-class relationships with his motivic material. The analysis of the first movement discusses its form, and regards a number of sections of the composition, showing the use of inversional symmetry, contrapuntal strategies, and the relation of contrapuntal and metric structure. The final section considers the issue of hypermeter in the first movement.

Honegger's second symphony, completed in 1941, arose out of an earlier commission by Paul Sacher. In a letter of June 11, 1936, Sacher had asked Honegger for a work for strings to commemorate the tenth anniversary of the Basel Chamber Orchestra. The celebration prompted several works by other composers, including Bartók's *Music for Strings, Percussion, and Celeste*, but Honegger's commitments prevented him from completing more than a short 56 measure *Largo* for string orchestra in September of that year.[1]

Honegger began sketching the symphony in November 1940. He later wrote:

> I have a very good friend in Basel, the conductor Paul Sacher. For nearly thirty years he has performed my works extensively and it is for him that I have written various new works. In 1938 he asked me to write a symphony for the *Basel Chamber Orchestra* or the *Collegium Musicum of Zürich*, both of which he conducts. For over a year I gathered together a few vague ideas; none of them satisfied me. Eventually during the sad days of the Occupation, I threw myself into the quartets of Beethoven and the influence of these magnificent pieces stimulated and enabled me to start work. I sent the score to Paul Sacher who gave it its first performance in Zürich on 18 May 1942.[2]

---

[1] Currently unpublished, listed as H. 105 in Harry Halbreich, *Arthur Honegger: Un musicien dans la cité des hommes* (Paris: Librairie Arthème Fayard, 1992), 439.

[2] Quoted and translated in Geoffrey Spratt, *The Music of Arthur Honegger* (Cork: Cork University Press, 1987), 395-6.

The second movement was completed first:

> ... for me, it is always by that central part of a triptych that the composition of a symphonic work begins, since I am only able, in exceptional cases, to conceive of it in other than tripartite form.[3]

This movement bears the date March 15, 1941. The first movement was finished May 8, 1941, and the third movement was completed last on October 13, 1941.

The score was sent indirectly to Sacher. During the early days of the Nazi occupation of Paris, Honegger was not allowed to travel to Switzerland, but his duties as critic for the journal *Comoedia* brought him to Vienna for a Mozart Festival toward the end of 1941. There he gave a photocopy of the score to the conductor Franz von Hoesslin, who was able to bring it into Switzerland. A second photocopy of the score was given to Alfred Schlee, the director of Universal Edition. The inability of Universal to print the parts on time forced a delay of the premiere from January until May of 1942.

The symphony closes with the addition to the strings of a trumpet chorale melody in D major. Given the circumstances of the Occupation during which the symphony was written, the tonal resolution at the end of the work encouraged later biographers and critics to interpret the composition programmatically. Speaking of the chorale melody, Pierre Meylan's opinion is typical:

> This unexpected appearance precisely signifies the deeper sense of the work. Without the Bachlike chorale, the significance of the work would remain unclear or limited. After both of the first movements the chorale has a symbolic significance and offers to this hopeless and desperate music a glimmer of hope. The individual who has cast off the yoke of tyranny and is true to himself is permitted to believe in the coming of a better time. The second symphony, which has been played in all the concert halls of the world, became understood as a message to people oppressed and to those whose individual freedom is threatened.[4]

---

[3] "...(C')est toujours par cette partie centrale du tryptique que commence, pour moi, la composition d'une oeuvre symphonique, de même que je ne puis que par exception la concevoir autrement que sous cette forme tri-partite." Arthur Honegger, *Ecrits*, Textes réunis et annoteés par Huguette Calmel (Paris: Librairie Honoré Champion, 1992), 171-2. Originally from *Mitteilungen des BKO* (Basel Chamber Orchestra), 9 October 1943.

[4] Pierre Meylan, *Arthur Honegger: Humanitäre Botschaft der Musik* (Frauenfeld and Stuttgart: Verlag Huber, 1970), 152. "Doch deutet gerade diese unerwartete Erscheinung auf den tieferen Sinn des Werkes hin. Ohne sie, ohne den Bach-ähnlichen Choral, würde die Bedeutung des Werks unklar oder beschränkt bleiben. Der Choral hat nach den beiden ersten Sätzen symbolische Bedeutung und setzt dieser hoffnungslosen und verzweifelten Musik einen Hoffnungsschimmer auf. Der Mensch, der das Joch der Tyrannei abgeschüttelt und sich selbst gefunden hat, darf an das Kommen besserer Zeiten glauben. Die zweite Sinfonie, die in allen Konzertsälen der Welt gespielt worden ist, wurde als Botschaft an die unterdrückten Völker und an die in ihrer persönlichen Freiheit bedrohten Menschen verstanden."

Another commentator places the work in the tradition of the "message" symphony developed in the early part of the century by D'Indy, Widor, and others, in which the symphonic program represents the symbolic triumph of light over dark, hope over despair, or good over evil.[5]

In a program note supplied to the Zürich journal *Alte und neue Musik*, however, Honegger disavowed any programmatic intent to the work, instead stating:

> My general preoccupations concerning this symphony are the same as those which dominate all my other symphonic works:
>
> 1. A rigorous form, a suppression of the recapitulation such as is found in Classical works where it always has the feeling of tedium.
>
> 2. A quest for themes which have sufficient character to attract the listener's attention and to enable him to follow the development of the whole story.
>
> 3. I have not searched for any program, nor a literary or philosophic foundation [*donée littéraire ou philosophique*]. If this work expresses or shows emotions, these come forward naturally, since I express my thoughts only in music and, perhaps, without being entirely conscious of them.[6]

As with *Pacific 231*, Honegger denied a programmatic foundation to the work, instead emphasizing formal and thematic considerations. He even suggested that he used the trumpet at the close of the symphony for orchestrational, not programmatic purposes, claiming that "it is not a desired effect, merely a support for the melody which is in dotted quarter notes in the first violins and which therefore runs the risk of being absorbed by the polyphony of the other instruments of the same timbre."[7]

Sketches for the Symphony

The compositional sketches for the symphony are contained within the third of Honegger's four sketchbooks held at the Bibliothèque Nationale in Paris. The first sketch begins on page 30. However, pp. 44-56 contain the body of the work. In general, Honegger brackets the 20-stave manuscript paper two staves at a time, writing in bass and treble clef. Although the second movement was

---

5 Brian Jack Hart, "The Symphony in Theory and Practice in France, 1900-1914" (Ph.D. diss., Indiana University, 1994). Hart also places Honegger's third symphony in this category: this symphony does contain an explicit program provided by the composer.

6 Translated into French and reprinted in Willy Tappolet, *Arthur Honegger* (Neuchâtel: Editions de la Baconnière, 1957), 239. English translation of numbers 1 and 2 from Spratt, *Arthur Honegger*, 396.

7 From the *Mitteilungen des BKO*; translated in Spratt, *Arthur Honegger*, 423.

completed first, large portions of the first two movements were written in alternation, and the first pages of the sketches often shift between movements. Figure 5.1 gives the order of material for the first two movements in the sketchbook, corresponding to the movement and measure number in the printed score.

Figure 5.1: *Symphonie pour cordes*, Order of composition from sketchbook

Sketchbook

P. 44

Movement	Measure
2	45-47
1	30-34 (these measures correspond to mm. 1-4 of the Allegro)
2	29, 33, 35
2	42-49
2	68-74
1	34-37
2	49-52
1	30-33 (alternative, not used in composition)

P. 45

2	1-24, 53-60
2	25-42

P. 46

1	117-122
1	86-93 (alternative)
1	94-96
1	9 extra measures
1	107-108
1	84-85
1	3 extra measures
1	114
1	90-102
1	106-117
1	253-254
1	103-105
1	257-258 (alternative)

P. 47

2	61-68

1	38-90
1	231-233 (with additional counterpoint sketches)

P. 48

1	240-244
1	143-149
1	214-245

P. 49

1	121-213

P. 50

1	253-264
1	268-271

P. 51

1	245-252

The pace of composition in the sketches clearly shows a contrast between shorter fragments in the initial tentative stages and longer sections once the movement takes shape. For example, fragmentary portions of the second movement appear on the first page of the sketches (p. 44), but by the next page the body of the composition has crystallized. Similarly, the lengths of passages in the first movement become gradually longer as the sketches progress. Honegger discusses this compositional process in *Je suis compositeur*:

> I first seek the contour, the general aspect of the work. Let us say, for example, that I see taking form a sort of palace in a very opaque fog. Reflection progressively dissipates this fog and permits me to see a bit more clearly into it. Sometimes a beam of sunlight appears to illuminate a wing of the palace as it grows: this fragment then becomes my model. When this phenomenon spreads, I begin my search for my materials for construction. I prospect in my notebooks.

Of the process of sketching and developing his ideas from his notebooks he continues:

> Gedalge gave me that habit. When a motif, a rhythm, an entire phrase comes to me, I jot it down. You know Beethoven's notebooks? Without being so presumptuous as to suggest a comparison, I admit that I proceed in the same way, and I recommend it to my pupils. So I check my notebooks of sketches in the hope of uncovering there some melodic design, a rhythmic plan, or some sequence of chords susceptible of being useful. Sometimes I think I have found what I am looking for,

Example 5.1: Chromatic expansion (sketches)

central movement of the composition but with the center of that movement.[11] The motive is transposed and shifted rhythmically in the composition. This alteration follows in the sketches two staves below, in the final four measures of an eight-measure section. Example 5.2a contains the material from the sketch, and Example 5.2b the material from the composition. The initial motive from the sketch is labeled Motive X. Note that the general process of chromatic expansion fostered in Example 5.1 is likewise present in Motive X.

Example 5.2: Motive X, sketches and score

[11] In two separate analyses, Halbreich points to Honegger's use of the golden section proportion, first in the discussion of the early *Le Chant de Nigamon* (1917), second in the third movement of the fifth symphony (1950), one of Honegger's last works. See Halbreich, *Arthur Honegger*, trans. Nichols, pp. 336 and 347.

In Example 5.2b, from the completed composition, the procedure of chromatic expansion is present in the outer voices of the four measures that precede the X motive. The material stated by the bass (and cello), clearly related to the X motive, moves chromatically upward from G2 to Bb2 within the first measure, and down to F#2 in the second measure. On the downbeat of the third measure the lower voices further expand the intervalic space by widening in both directions to the F2-B2 harmonic interval.

In the upper voice of 5.2b, the opening pitches of C6, Db6, and Eb6 form an [013] set class. The Eb completes the chromatic aggregate in this passage. Moreover, the first two pitches of this upper voice motive (C and Db) anticipate the beginning pitches of the X motive which follows, and the following X Motive widens the intervalic boundary from three semitones (C-Eb) to four (C-E).

In Example 5.2b, the use of the identical pitch classes of C and Db to begin both the upper-voice melody (first measure) as well as the X Motive (third measure) indicates that this stage of the compositional process moves from relationships based upon an inchoate notion of chromatic expansion to one which establishes pitch-class relationships. Subsequently in the sketches, the motivic relationship becomes based upon pitch; this appears in Example 5.3. The end of the X motive, corresponding to mm. 47-48 of the printed score, is followed in the sketch by the final seven measures of the composition, mm. 68-74. Example 5.3 shows that these pitches which complete the X motive, filling in the chromatic space from E4 to C#4, are echoed by the identical pitches in the following melody in the sketches. Now the general notion of chromatic expansion is confined to an intervalic identity of three semitones, and both motives of Example 5.3 assert not only a pitch-class but a pitch identity. Although adjacent in the sketches, in the completed composition this pitch relationship extends across twenty measures, occurring first at mm. 47-48, and returning at mm. 68-70.

Example 5.3: End of X Motive and final melody of mvmt. 2 (sketches)

Similarly, directly beneath the X Motive in the sketches occurs a melody whose opening three pitches state the motion from E6 to C#6,[12] with the beginning E metrically placed in the identical place as that above. Example 5.4 places these two motives together. The motive following X retains the same pitch classes, although the move to C# forms an [013] motive, rather than filling in the chromatic space. While the E-C# pitch class relationship between both motives is similar to other relationships discussed, Honegger discards this pitch-class relationship. As implied by an indication "1/2 plus haut" in the margin of the sketches, its m. 49 appearance in the score, directly following the X motive, is a half-step higher. Example 5.5 shows the two motives as they appear in the published composition.

Example 5.4: End of X Motive and following section (sketches)

Example 5.5: End of X Motive and following section (score)

Is the replacement of a pitch-class relationship with the move a half-step higher a voice-leading maneuver? As the dotted slurs in Example 5.5 imply, the half-step transformation allows the E-C# intervalic boundary formed by the final pitches of the X motive (m. 48) to be melodically answered by that of F-D at m. 49; these pitch classes also form the outer pitches of the accompanimental harmony on the downbeat of m. 49. M. 49, then, expresses the D/F intervalic boundary both vertically and horizontally. Moreover, in the composition this

[12] This melody continues through to A#5, and is symmetrical around C#6.

intervalic shift coincides and underlines the shift in texture, and the harmonic climate now becomes diatonic.

The final measures of the composition (mm. 68-74), included in their entirety on the first page of sketches, reiterate this particular intervalic shift more clearly. The score appears in Example 5.6a. In order to show these intervalic shifts more clearly, a textural simplification is given directly below. The E-C# melody is given to the cello solo between mm. 68-70; the pitch-classes are echoed, in contrary motion, in the bass at m. 70. In m. 71, the bass fills in this intervalic span a half-step higher. The move from F2 to D2 in the bass responds to the E4 to C#4 melody within the first three measures. As in Example 5.5, the relationship departs from pitch-class associations, instead developing a process that shifts the intervalic span upward a half step.

Moreover, Example 5.6b shows this process replicated in an inner voice in m. 72, the antepenultimate measure. Once the bass has fallen to the final pitch D, the melodic motion of B2 to G#2 in the cellos is followed by that from C2 to A2. Thus the initial half-step motion of the ic3 boundary which occupies the entire seven-measure phrase is repeated, nested within the final three measures of the composition.[13]

This procedure differs from that used in earlier motivic relationships, which previously were associative, based upon pitch and pitch-class return. The T1 shift of an intervalic span introduces an element of process to these previously static recurrences. While these intervalic shifts sometimes take place across different textural or registral strata, they provide a cadential and articulative role within the composition.

Sketches and Published Composition, Second Movement

The body of the composition is already completed by the second page of the sketchbook. While little is altered from the sketch to the composition, the strophic and additive structure of the repeating ground section is demonstrated by Honegger's use of six staves for this section in the sketches. The lower two

[13] The final notes played above the D-A bass are Ab^3 and F^3. While this does not exemplify the processive relationships discussed, it does confirm the importance of this interval which has formed the intervalic boundary of many of the motivic relations discussed. Spratt calls general attention to the importance of this interval in the B section: "the material of this part of the movement is equally dominated by the interval of a third—in particular the minor form," but does not discuss where or how this interval is deployed. Other writers discuss "dominant" intervals in the work: Halbreich mentions the "predilection" for tritones and minor seconds; Maillard and Nahoum speak of the harmonic intervals of the major seventh, minor ninth, and diminished octave. See Spratt, *Arthur Honegger*, 399; Halbreich, *Arthur Honegger*, trans. Roger Nichols, 312; and Maillard and Nahoum, *Les Symphonies d'Arthur Honegger*, 64.

Example 5.6a: Mm. 68-74, end of mvmt. 2

Example 5.6b: Reduction of mm. 68-74

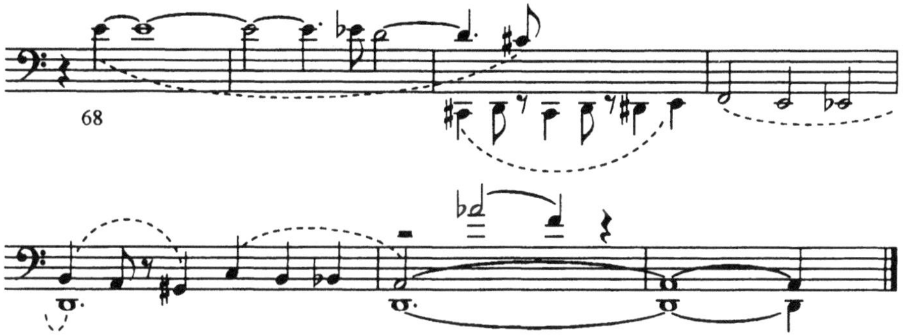

staves contain the ground. Directly above, the cello melody—stated with the second and third statement of the ground (mm. 9-16, 17-24)—takes up the third stave. The violin melody stated with the third ground statement (mm. 17-24) occupies the upper (sixth) stave, and the violin melody of the final A section (mm. 53-60) is given in the fifth stave. The fourth stave contains the alterations to the ground bass that occur at mm. 56 and 60 of the composition.

The opening eight measures of the movement are given in Example 5.7. The semitonal voice-leading of the repeated ground reflects the notion of chromatic expansion found in the early sketches. Honegger develops this further here, circulating the aggregate, so that the sole missing pitch between mm. 1-4, F, is stated in the violas on the downbeat of m. 5. Additionally, the intervalic range, especially in the upper two voices, expands in the second four measures. The intervalic range of violin 2 moves from four semitones (mm. 1-4: Db4- A3) to nine (mm. 5-9: F#4- A3), while in the violas the range moves from four (F#3-Bb3) to eight (C4- E3).

The B section is begun directly below in the sketches, and the twenty-six measure section contains a chromatic bass rising from C to F# before reaching G, which precedes the statement of the X motive at the golden section. Geoffrey Spratt, writing of this chromatic ascent, suggests that it is "prevented from being an embarrassingly weak device which Honegger uses all too often by the irregularity of its periodicity."[14] Yet the periodicity of the chromatic climb is not completely irregular, and the regularities offer some relationship to the motivic identities and intervalic boundaries developed in the earlier sketches. Figure 5.2 gives the length in measures for each pitch in the bass for the chromatic ascent of the B section.

---

[14] Spratt, *Arthur Honegger*, 399. Honegger uses a similar technique in the second movement of his first symphony.

Example 5.7: Mm. 1-8, second mvmt.

Figure 5.2: B section from second mvmt.

Pitch:	C	C#	D	Eb	E	F	F#
Length:	1	2	1	4	2	1	4
Measure:	25	26	28	29	33	35	36

Figure 5.2 demonstrates that, save for the initial bass pitch of C, the lengths are periodic at three semitones. This partition of the chromatic space then resonates with the surface motivic associations worked out in the first page of the sketches, which develop motives that articulate this span of three semitones. Honegger organizes this ic3 division of the chromatic space by weighting the arrivals on Eb (m. 29) and F# (m. 36) with the extended length of four measures. In addition, motivic and harmonic material recurs at T3 at m. 29 and m. 36. Example 5.8 displays this recurrent material: the harmony in the inner voices articulates an [012] pitch-class set at mm. 29 and 35. In both cases, this is preceded by [016], which appears at mm. 28 and 34. In addition, above the bass pitch of Eb (m. 29) and F# (m. 36) is repeated a melody that is inversionally symmetrical: C4-E-F-Gb-Bb (symmetrical around F) at m. 29 is then restated at m. 35 at T3 (Eb4-G-Ab-A-C#5, symmetrical around Ab). Thus, this transpositional level of three semitones reflects the move in the bass from Eb to F#.

Example 5.8: Chromatic ascent of B section

Moreover, this ic3 relationship is duplicated on different levels of the texture. The melody, beginning at m. 30, contains a retrograde form of the chromatic bass ascent from C to F#. As the dotted slurs in mm. 30-33 of Example 5.8 indicate, the melody descends chromatically from Gb5 to C. As in the bass ascent, the chromatic space again is partitioned by three semitones, with Gb, Eb, and C durationally weighted in relation to the other chromatic pitches of the melody.

Hardly a near-prosaic and simplistic device, the ascending chromatic bass in the B section is structured around pitches that resonate with the abstract intervalic ideas developed in the earliest stages of composition. Durational weight is given to divisions of the chromatic space at the ascending interval of three semitones in the bass, and this periodicity is reinforced by recurring melodic and harmonic events at different levels of structure. We have seen that the ic3 boundary plays a role in the X Motive, as well as in other motives indicated in Examples 5.2-5.6. The surface motives initiated in the first page of the sketchbook provide the large-scale organization of the B section, forming focal points of division within that section.

First Movement: Form

Regarding the first movement of the symphony, Honegger writes "I worked for a long time on the first movement in order to give it a form which was sufficiently concentrated and rigorous without destroying the interior violence which I am proud to say I have managed to create and which counteracts the austerity of the initial motif."[15]

Despite its rigor, the form of the first movement has been the object of some debate. Both Halbreich and Spratt consider the movement to be in sonata form, although their analyses differ in identifying the point of recapitulation. Parker, too, labels the movement as "modified sonata form," and the thematic locations differ from Halbreich and Spratt. Landowski describes the movement as an "asymmetrical rondo," while Maillard and Nahoum present the form of the movement as hybrid, combining sonata form with a rondo that includes four refrains with three couplets.[16]

The movement's formal complexity arises from several sources. The introductory section (Molto moderato, mm. 1-29), characterized by a repeated

---

[15] Quoted and translated in Spratt, *Arthur Honegger*, 406.

[16] See Spratt, *Arthur Honegger*, 406; Halbreich, *Honegger*, trans. Roger Nichols, 312-14; Daniel Quienton Parker III, "The Symphonies of Arthur Honegger: An Examination of Stylistic Factors and Form" (Ph.D. diss., University of Kentucky, 1990), 413; Marcel Landowski, with the collaboration of de Marie Françoise Christout, *Arthur Honegger* (Paris: Editions de Seuil, 1957), 135; and Maillard and Nahoum, *Les Symphonies d'Arthur Honegger*, 58.

two-measure motive in the violas (Motive α), returns twice: first at mm. 154-175, and again at mm. 253-267. In addition, the recapitulation remains highly developmental, and Honegger alters the order of the themes relative to the exposition. The following provides a formal summary of the movement.

EXPOSITION	
Introduction (with viola motive α)	1
Allegro	
Theme 1	30
Theme 2 (introduces triplet motive β)	54
Un poco meno mosso	
Theme 3	70
a Tempo	
Development of motives α and β	78
α motive and Theme 1 material (cellos/basses)	97
DEVELOPMENT	
Material from Theme 1 and α	105
RECAPITULATION	
Molto Moderato	
Reprise of Introduction	154
Allegro	
Theme 2	176
Poco meno mosso	
Theme 3 and opening of Theme 1 (in double canon)	190
Tempo	
Development of Theme 1 and β	200
Development of Theme 1 and α (in Violins 2 and in augmentation Violins 1)	229
Molto Moderato	
Reprise of Introduction with Theme 3 (Violins 1)	253
Allegro	
Theme 1	268

Although the composition opens with a twenty-nine measure slow introduction, Honegger began the sketches of the first movement with the two four-measure unison motives that initiates the Allegro section at m. 30. The two four-measure phrases, creating a parallel period, are given in Example 5.9. With the unison theme, Honegger creates aggregate or near-aggregate completion: the first four measures lack F# and E to complete the aggregate, and F# appears as the first note of m. 5.[17]

Example 5.9: Theme 1 (first movement)

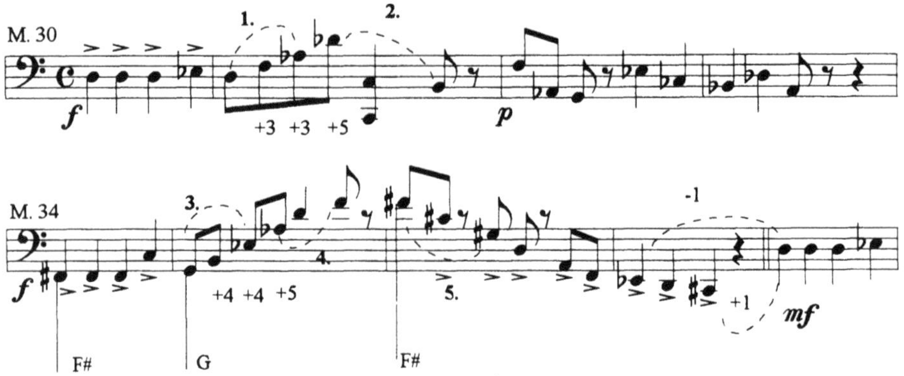

Despite this almost complete circulation of each of the pitch-classes of the aggregate, the first measure asserts pitch centricity through repetition of the pitch D, as well as through a half-step departure and return gesture (D-Eb-D) through the downbeat of the second measure. Example 5.9 shows that the following four measures mirror this half-step departure and return on a higher level of rhythmic structure, asserting the pitches F#-G-F# on the successive downbeats of mm. 34-36. In addition, through inversional symmetry around D, the pitch content of m. 37 (Eb-D-C#) brings about a return to the unison D and a restatement of the theme in the following measure.

Although the rhythmic relationship of mm. 30-31 and mm. 34-35 is parallel, the pitch content varies. However, the relationship between m. 31 and m. 35 establishes an expansion of intervalic content, in which the directed interval content at the beginning of m. 31, <+3 +3 +5>, becomes expanded to <+4 +4 +5> in m. 35. In addition, segments of the theme are developed through interval cycle content. The interval cycle of 3 stated at the beginning of measure 31 (this is indicated as "1" above the score in Example 5.9, and marked with dotted slurs) is followed by the projection of an interval cycle of 1 with the remaining pitches in the measure, Db, C, and B (indicated as "2" and marked with dotted slurs). At m. 35, the interval cycle of four (#3) is succeeded by three pitches of interval cycle 3 (#4)—Ab, D, and F. Finally a descending form of an interval cycle of 5 (#5) follows with the pitches F#, C#, and G#.[18]

---

[17] The E appears in the twelfth measure, in the stepwise descent of the cellos.

[18] Additionally, the content of m. 34 suggests an orientation based on ic 6. Laurent Fichet briefly discusses the notion of thematic content generated by constituent intervals in

However, thematic identity in later appearances of the theme is not based upon intervalic content. Example 5.10 contains two restatements of the opening measures of the theme, the first from mm. 38-41, the second from mm. 50-53. While the intervalic disposition of the theme is altered, rhythm and, frequently, contour are retained. Thus, the dimensions of rhythm and contour provide the foundation for thematic identity, while intervalic identity is discarded.[19]

Example 5.10: Theme 1 restatements

Inversional Symmetry and Directed Motion

The movement employs inversional symmetry, developing its potential in several ways. For example, inversionally symmetrical relationships of the outer voices may structure the harmonic dimension. Second, inversional symmetry can be projected in the melodic dimension, creating directed linear motion in a single melodic voice. Third, inversional symmetry may engender pitch centricity through motion toward a focal central pitch. Finally, inversional symmetry may generate longer-range directed motion as well.[20]

The four measures preceding the m. 50 return of the first theme provide examples of several of these techniques of inversional symmetry. Example 5.11 contains these measures. Between mm. 46-47 symmetry is veiled to some extent. The motion from G4 to E4 in the upper part is asserted twice: the first instance moves through F natural, with the directed interval content <-2 -1>. The second instance progresses through F# with the directed interval content <-1

"Problèmes de langage dans l'oeuvre symphonique d'Arthur Honegger," in *Honegger-Milhaud: Musique et Esthétique* (Paris: Librairie J. Vrin, 1994), 119.

[19] However, the <+4 +4 +5> from m. 35 of the original theme (Example 9) is rearticulated in m. 39, and again in m. 53.

[20] Chapter 3 presents in some detail these issues related to inversional symmetry. Some of the discussion here uses the sums of pitch classes dyads (with C = 0, C# = 1 ... Bb = 10, and B = 11) in order to show inversional symmetry.

-2>. Beneath, the bass rises twice to B3, both times through G#. This establishes a veiled sum 3 voice relationship, shown by the diagonal lines in the score, associating G# with G and B with E. Additionally, as Figure 5.3 indicates, the directed interval content for these two measures in the bass *reverses* that of the upper part: at m. 46, <-2 -1> (upper voice) is stated against <+1 +2> (lower voice); at m. 47, <-1 -2> is stated against <+2 +1>.

Figure 5.3: Directed interval content in mm. 46-47

Violins	G	F	E	G	F#	E
		-2	-1	+3	-1	-2
Basses	G#	A	B	G#	A#	B
		+1	+2	-3	+2	+1

Example 5.11: Inversional symmetry at m. 46 ff.

The outer-voice motion of the following two measures shifts the inversional center upward to sum 4 with the dyads of C-E and Db-D#. Unlike the previous sum 3 voice relationships, these dyads are stated simultaneously. The dyads suggest a convergence upon D. Instead, however, both voices fall to the unison B.

Yet not only is D absent, but the shift of both parts to B prevents *its* sum 4 inversional partner, F, from being stated. However, as the passage continues, it enacts directed motion by yielding both these two pitches and placing them in registral relief. The "missing" D becomes the highest pitch of the following two measures. In addition, the "missing" inversional partner of B, F, becomes the highest pitch of the following two-measure phrase.

Within this passage, then, inversional symmetry controls the outer-voice relationships, and this is projected in multiple ways. In mm. 46-47 the symmetry is veiled, with the inversionally related voices given differing rhythmic identities. Inversional symmetry is asserted overtly with the simultaneous contrary motion of mm. 48-49. Finally, the absent inversional partners of D (mm. 48-49) and F (m. 50) provide a sense of directed motion to the mm. 50-53 return of the theme, where these pitches become subsequently highlighted at m. 51 and m. 53.

Contrapuntal Stratagems

Throughout the movement, much of the source of musical development comes from the working out of contrapuntal possibilities of motives derived from the initial eight-measure theme. Example 5.12 shows mm. 107-120, which present the last five pitches of Theme 1 in a series of stretti. Of the eleven entries, the first eight alternate with a statement in inversion, and the final three statements state simultaneously the motive with its inversion. Note, too, that beginning with the third entry, each entry is accompanied by a chromatic "shadow" line, following the direction of the motive. Beneath the example, Figure 5.4 gives the beginning pitch for each statement, the interval class distance ("Ic") for each statement, and the durational interval ("Dur. int.") in eighth notes between each statement.

The durational interval row indicates a general accelerando, followed by a decelerando in the last three entries. Only the second, third, sixth, and eighth entry corroborate the written meter of 4/4 with entries at four and two eighth notes. Within the passage there are some correspondences of interval-class to durational interval. For example, both entries at the interval class of 2 are given the durational interval of 4, and both entries of interval class of 1 are given the durational interval of 2.

In addition, the final three entries pair the interval class of three with the durational interval of 5, and the sequence here, while contradicting the notated meter, serves to direct the move back to D and the recurrence of the initial theme. Within the T3 ascent, the move to D is further enhanced in the upper voice through an ic1 ascent. The upward stems in Example 5.13 show that the upper registral extreme of these four consecutive entries asserts a chromatic ascent from E5 through D6.

Example 5.13 also shows that the following measures continue abstractly the chromatic ascent. Here the inversional center of the registral boundaries of each measure systematically ascends with each measure. This is indicated in the lowest system, which shows the boundary intervals for each measure, along with

Example 5.12: Mm. 105-117

Example 5.12 (Continued)

Figure 5.4: Mm. 107-116

Pitch	B	Db	B	B	E	F	B	C	Eb	F#	A
Ic		2	2	0	5	1	6	1	3	3	3
Dur int.		4	4	3	3	2	3	2	5	5	5
Measure	107	108	109	110	111	111	112	112	114	115	116
Instrument	Bass	Vla	Cllo	Vn2	Cllo	Vn1	Vla	Vn1	Vn	Vn	Vn

the inversional center for these boundary intervals. For example, the registral extrema of m. 118, Db5 and B3, imply an inversional center of F#^{4} (sum 0). The next measure shifts upward, implying an inversional center of F#-G (sum 1), and the following measure the inversional center shifts upward to G (sum 2).

Example 5.13: Ic 1 ascent, mm. 112-120

Mm. 85-97: Elaboration of Simpler Background

The strategy of articulating arrival points through chromatic ascent or through inversional symmetry is corroborated by the sketches. In fact, the sketches reveal that certain approaches to structural points are calculated through a simpler background; this background then becomes elaborated in the composition. Example 5.14 is reproduced from the sketches and shows Honegger plotting the A6 arrival at m. 94 through chromatic ascent initiated from F5. The chromatic motion in the upper voice is added onto a vertical harmonic framework, formed out of triadic pairs whose roots are predominantly a step apart. This vertical texture is altered with the arrival on A, which establishes the repeated downward arpeggio of an F# minor triad, implying a layer of 3/4 meter. This arrival on A6 corresponds to m. 94 of the published composition.

Example 5.14: Chromatic ascent, sketches

To a certain extent, the published composition camouflages the obvious and prosaic chromatic ascent of the sketches. Instead, the ascent is asserted irregularly, but given important metric or registral placement, and alterations are made to other sketch passages in order to accommodate this ascent. For example, 5.15a gives the initial sketches for mm. 85-86. Here the passage restates an [013] motivic cell in the first violins, appearing as B5-A-G#. A stepwise ascent (F, Gb, Ab, Bb) precedes the statement of the cell a half-step higher, C-Bb-A.[21] Example 5.15b shows these measures as they appear later in the sketches and in the composition. The entire passage is extended, and a pitch alteration takes place in m. 85. In comparison with the sketch of Example 5.15a, the stepwise ascent of 5.15b is extended, and the [013] cell occurs on the downbeat of m. 86, beginning on F#. The lengthened stems in Example 5.15b show how this elaborated version launches the beginning of a protracted ic1 ascent, connecting E# and F# at the downbeats of mm. 85-86. This corresponds to the beginning of the overt chromatic ascent in the sketches, shown in Example 5.14, in which these two pitches begin the chromatic ascent en route to A.

The score given in Example 5.16 indicates that this alteration allows a surreptitious projection of the sketch's patent chromatic ascent: E#5 to F#6 are connected on the downbeats of mm. 85-86. Following, the connecting chromatic pitches are asserted in the upper register: G6 is stated in the next measure at the upper registral extreme, while G# is withheld until m. 92, again asserted in the

[21] This [013] melodic cell, with the directed interval content of <-2 -1>, recurs throughout the movement, first appearing as the viola solo in the slow introduction. As Spratt claims, this cell "dominates the entire movement like a motto." Spratt, *Arthur Honegger*, 406.

Example 5.15: Mm. 81-85, sketches and score

upper register before reaching A in m. 94.[22] Thus Honegger works out the prosaic chromatic ascent from F to A (indicated in the sketches in Example 5.14) in a more complex and sophisticated fashion.

In addition, the overt ic1 motion of the sketches is transferred to the entire structure. Each of the parts establishes two-measure cells that repeat at T1 from m. 90 through the downbeat of m. 94. Example 5.17 contains a reduction of these measures: Violins 2 progress from E (m. 90), F (m. 92) to F# (m. 94), while violas chromatically connect G to A. With few exceptions, the outer voices likewise participate in this T1 ascent between mm. 90-94.[23]

This projects the melodic chromatic ascent to the complete structure in these measures. This ic1 motion that brings about the arrival at m. 94 is maintained afterward. The implied layer of 3/4 is retained from the earlier sketch and stated in imitation in the basses one beat later. Yet the repeated arpeggio of the F# minor triad from the sketches is now altered, reflecting the chromatic shifts heard previously. The lower pitch of the three-note arpeggio takes on this motion, moving chromatically from C# through D# during these three measures. This chromatic motion is shown in Example 5.17 by the stemmed pitches in mm. 94-96.

The composition thus elaborates the simple F-A chromatic ascent of the original sketches (shown in Example 5.14), masking the overt aspects of this

---

22 The first violins also overreach the A by a half step, stating Bb in m. 89.

23 These outer voice exceptions occur upon the downbeats of mm. 90 and 92; in addition, the basses' final quarter note of D at m. 90 is repeated in the analogous place in m. 92, avoiding an exact T1 relationship.

Example 5.16: Mm. 85-96

Example 5.16 (Continued)

Example 5.17: Reduction of mm. 90-96

ascent through metric and registral placement, taking place in the composition at mm. 84-94. In addition, this chromatic ascent is transferred to different aspects of the structure, shown by the given harmonies in Example 5.17 (mm. 90, 92, and 94), heightening the arrival at m. 94 with the implied 3/4 meter. Additionally, chromatic motion continues further after this m. 94 arrival, infiltrating the compound melody at m. 94-96, linking C#-D-D#.

Thematic Restatement: Honegger's "Melodic Balances"

This section at mm. 94-96, with its implied meter of 3/4, is eventually followed by a restatement of Theme 1 on C# at m. 105 ff., and another statement of the theme on D at m. 117 ff. In considering the significance of C# for the m. 105 return to the first theme, several possible rationales may be offered. In a general sense, perhaps, thematic return at the transpositional interval of a semitone is favored by the composer.[24] In another sense, the C#-D motion may work for the composer in analogy to a longer-range composing out of leading-tone motion.[25]

However, in his prose writings, Honegger is explicit about the rejection of the use of pre-determined tonal schemes, giving renewed emphasis instead to unity provided by "the totality of melodic and rhythmic relationships," and placing greater importance on "melodic and especially rhythmic balances" than on "tonal equilibrium."[26] This may suggest that it is more fruitful to investigate focal pitch arrival points and relationships as emerging more spontaneously from melodic relationships than from a precompositional tonal scheme. Honegger's own compositional dicta seem to suggest exploring mid- and local-level melodic relationships rather than global tonal relationships.

Therefore, the implied 3/4 passage of mm. 94-96 and the following transitional passage at mm. 97-104 may provide some clues to the eventual arrival on C# at m. 105. As the stemmed pitches in Example 5.17 show, at mm.

[24] Elsewhere, the section of mm. 94-96 recurs a half-step lower, in mm. 226-228.
[25] This provides the same strategy as in *Rugby*. See Chapter four.
[26] Honegger, *I am a Composer*, 83.

94-96 the descending F# minor triad is altered to include a chromatic ascent from C# to D#. The ascent projects a moving aural image against the static ostinati in both the inner parts and the upper pitches A-F# of the descending arpeggio.

Example 5.18a contains the eight-measure passage that follows the section at mm. 94-96. The passage establishes a melodic ostinato within mm. 97-104, and repeats the [013] cell in the upper voice every two measures, falling an octave in mm. 101-102, and another octave the following two measures in mm. 103-104. This melodic ostinato presents a fixed structure against the moving part in the bass.

The bass is derived from the opening half-step motion in quarter notes from Theme 1. Here its pitch focus shifts every two measures, centering first around D# (mm. 97-99), then G# (99-100), before moving to C# for two measures prior to the return of the first theme on C#. In examining these moving parts, a subtle contrapuntal melodic relationship emerges, shown in Example 5.18b, which omits the ostinati from both sections at mm. 94-96 and 97-104. The chromatic motion upward from C# to D# in the upper voice in mm. 94-96 is answered in mm. 97-104 by the bass, progressing from D# to C#. The retrograde relationship is not exact, however. The bass part, rather than moving chromatically down, is twice projected down a perfect fifth to link the D# to C#. We can hear Honegger's ideal of "melodic balances" worked out here, and serving to structure locally the arrival on C# of the first theme at m. 105. The mm. 94-96 ascent from C# to D# is balanced by the descent from D# to C#.[27]

Contrapuntal Stratagems: Transpositional Levels and Metric Structure

Elsewhere, the sketches point out systematic contrapuntal development, some of which is not retained in the composition. From the published composition, Example 5.19 shows mm. 231-232. The four-note [0258] motive is successively given to each instrument at a pitch level of nine semitones higher than the previous statement, and imitated at the durational interval of the half note. The pitch-class content of the entire passage asserts the C#-D octatonic collection.[28]

The sketches contain other contrapuntal possibilities worked out with this motive, including imitation at other durational intervals. Example 5.20 is taken from p. 47 of the sketches, and shows three different passages based on the same [0258] motive. 20a displays imitation at the quarter note. Beside it in the sketches is 20b; here the imitation is at the eighth note. 20c presents imitation at both quarter-note and eighth-note durational intervals. Here the octatonic collection is

---

[27] In Violin 1, the chromatic motion at mm. 94-96 in fact progresses from C#-E: the D#-E motion is echoed in the bass in the following m. 97.

[28] This octatonic collection consists of the pitch classes C#, D, E, F, G, Ab, Bb, and B.

Example 5.18a: Mm. 97-106

Example 5.18a (Continued)

Example 5.18b: Mm. 94-106

Example 5.19: M. 229 ff.

asserted by statements at T3. In these instances, Honegger works out the contrapuntal possibilities at systematic durational intervals as well as at systematic transpositional levels.

In the composition, contrapuntal possibilities of other motives are worked out at other transpositional levels. Example 5.21a shows a passage at mm. 220-225 of the composition in which the transpositional level is ic4. The material in this passage states two motives contrapuntally, both derived from the initial eight-measure theme. One motive is taken from the seventh measure of the initial theme and stated in the upper parts. A second motive states material from the third and fourth measures of the initial theme in the lower parts. This passage precedes a recurrence of the implied 3/4 passage at mm. 226-228, heard earlier at mm. 94-96.

Example 5.21b shows the beginning pitch for each statement. In the passage, the T4 cycles are systematically played out. The motive in the lower parts recurs at the measure level, beginning on the pitches C (m. 220, cellos 2), E (m. 221, basses), and G# (m. 222, cellos 2), before the cycle repeats again at mm. 223-225, beginning in the basses. In the upper parts, the motive initially recurs at the measure level, beginning on A (m. 220, violins 1 and 2), C# (m. 221, violas), and F (m. 222, violins 1 and 2). This triple grouping is maintained in the following three measures. However, the upper voice motive now occurs in stretto at the half-note level, and the cycle is completed twice between mm. 223-225.

The systematic use of the ic4 as transpositional interval in this passage plays handily into the metric structure. Each completion of the T4 cycle occupies three measures, thus twice establishing a higher-level metric organization of three measures. This triple periodicity—the hypermeter of three measures—then anticipates the recapitulation of the implied 3/4 section that directly follows, which yields a triple grouping at the level of the beat. Here contrapuntal structure and metric structure are inextricably linked: the working out of ic4 transpositional cycles brings about the triple periodicity of mm. 220-225.

The appearance of the earlier implied 3/4 section at m. 94-96 of the exposition is likewise anticipated, albeit in different fashion.[29] In the exposition, hypermetric groupings of mm. 86-93 are quadruple, but the lower-level triple rhythmic subdivisions of the beat prepare the triple meter of the following implied 3/4 passage. In that instance, the gradual saturation of triplet subdivisions then project onto the 3/4 metric level at mm. 94-96. In the recapitulation, on the other hand, a higher rhythmic level—that of hypermeter—prepares the following implied 3/4 section at mm. 226-228. The transformational relationship of both sections is shown in Example 5.22.

---

29 This is discussed in further detail in Chapter 3.

Example 5.20: Contrapuntal sketches

Example 5.21a: M. 220 ff.

Example 5.21a (Continued)

Example 5.21b: Contrapuntal reduction

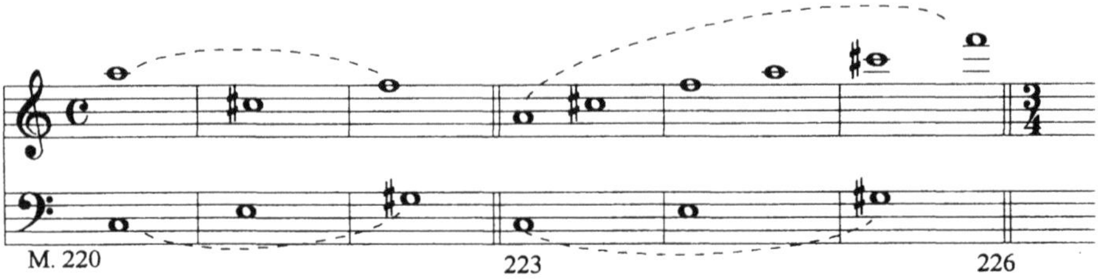

Example 5.22: Relationship between rhythmic levels, exposition and recapitulation

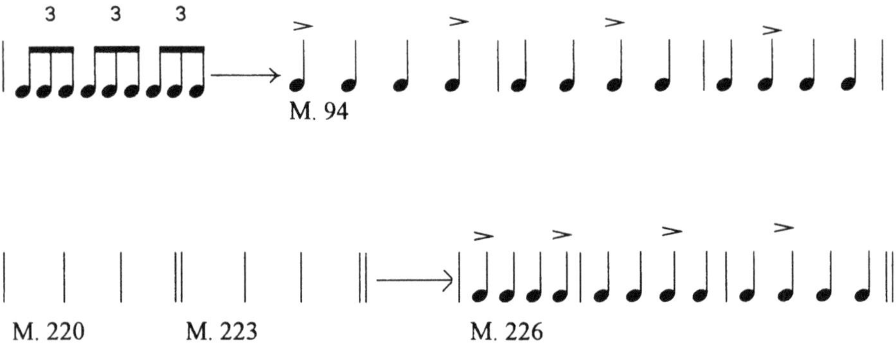

The comparison of these two passages illuminates Honegger's technique of creating relationships that straddle different levels of rhythmic and metric structure. In the passage from the recapitulation, we hear that these relationships operate in tandem with contrapuntal structure, and that systematic contrapuntal organization establishes and regulates metric structure at various levels. These contrapuntal processes participate in structuring the richly multi-leveled rhythmic relationships that take place.

Hypermeter in the First Movement

This final section examines the overall hypermetric organization of the first movement. The following three figures provide the overall sectional lengths within the exposition, development, and recapitulation. The lengths in measures for each of these sections reveal their large-scale metric organization. Following each of these three figures, the analysis examines the smaller subsections, which articulate the hypermeter of the movement. Figure 5.4 shows the large-scale metric organization of the exposition, following the slow introduction and beginning with the m. 30 Allegro.

Figure 5.4. Large-scale Metric Organization of the Exposition, *Symphonie pour cordes* (1st movement)

Section	Length of Section (in measures)
Theme 1 (m. 30)	24
Theme 2 (m. 54, R. 3)	16
Theme 3 (m. 70, R. 4)	8
α and β motives (m. 78, a tempo)	16 + 3
α and Theme 1 (m. 97, R. 6)	8

The metric organization of the sections within the exposition suggests a heavy reliance on regular groupings in multiples of four and eight measures. Even the seemingly irregular grouping of 3 measures appended to the 16-measure section at mm. 78-96 qualifies. Example 5.22 from the previous discussion showed that this notated three-measure section consisted of four bars of an expressed 3/4 meter.

Nevertheless, although the section lengths indicated in Figure 5.4 are primarily multiples of four- and eight-measure groupings, these sections do not necessarily parse into consistently regular four- and eight-bar hypermeasures. For example, the section at Theme 2 (mm. 54-69) consists of 16 measures, but the underlying grouping consists of five hypermeasures of 4 + 2 + 4 + 2 + 4 notated measures. (Moreover, some of the motivic material here masks these hypermetric divisions: the phrases in the cello cut across the division between mm. 59-60.)

In addition, the mm. 78-96 section (a tempo) consists of a grouping of 16 + 3. While the 16 measure section (mm. 78-93) forms an 8 + 8 grouping, these first eight measures at mm. 78-85 consist of three hypermeasures formed of 3 + 2 + 3 measures, working against an underlying four-bar hypermeter.

Figure 5.5 shows that the large-scale metric organization of the development section offers even less regularity than the exposition.

Figure 5.5. Large-scale Metric Organization of the Development Section, *Symphonie pour cordes* (1st movement)

Section	Length of Section (in measures)
M. 105	12 (4 + 8-bar contrapuntal extension)
M. 117 (R. 7)	10
M. 127	10

M. 137	8
M. 145 (R. 9)	9

Following the 12-bar opening consisting of a 4-measure statement of Theme 1 and a contrapuntal extension of 8 measures, there appear two consecutive 10-measure sections. Despite the similarity of their overall lengths, these two 10-bar sections are organized somewhat differently. At mm. 117-126, the organization consists of two hypermeasures of 4 + 6 notated measures, while the following section at mm. 127-136 has two hypermeasures of 5 (= 3 + 2) + 5 notated measures. The final 9-measure section at mm. 145-153 consists of a grouping of 5 + 4.

Figure 5.6 shows the large-scale metric organization for the recapitulation, following the return of the introduction at mm. 154-175.

Figure 5.6. Large-scale Metric Organization of the Recapitulation, *Symphonie pour cordes* (1st movement)

Section	Length of Section
Theme 2 (m. 176)	14
Theme 3 and 1 (m. 190, Poco meno mosso)	10
Theme 1 and β (m. 200, Tempo, R. 12)	11
Theme 1 and β (m. 211, R. 13)	9
Theme 1 (m. 220)	9
Theme 1 (m. 229, R. 14)	8
Theme 1 and α (m. 237, R. 15)	16
Introduction and Theme 3 (m. 253, Molto moderato)	15 (= 2-bar ostinato 6 times + 3 mm.)
Theme 1 (m. 268, Allegro)	17

Only the first four measures of the recapitulation statement of Theme 2 at m. 176 are based on the exposition statement of Theme 2 at m. 54 (R. 3) of the exposition. (In addition, the melodic theme in the recapitulation appears four semitones below the earlier melodic statement from the exposition. Moreover, the exposition section occupied 16 measures, while the corresponding section in the recapitulation occupies 14 measures.) Despite these differences, it is interesting to note an underlying metric correspondence to these sections. While Theme 2 at the exposition (m. 54) groups into 4 + 2 + 4 + 2 + 4, Theme 2 in the recapitulation (m. 176) groups into 4 + 2 + 4 + 2 + 2. Although only the first four measures have the same melodic material between these two sections, Honegger nevertheless duplicates the underlying metric grouping for the first twelve measures of each.

The 10-measure section at m. 190 (Poco meno mosso) is a double canon, using imitative entries of both Theme 3 (Violins 1 and Cellos) and material derived from Theme 1. Here Honegger elasticizes Theme 3 in relation to its earlier statement in the exposition (m. 70), adding a one-measure prefix to the theme at m. 190 (Violins 1). This prefix is also given to the Cellos in the following m. 191. As a result, the original eight-measure Theme 3 now becomes nine measures. The completion of this Theme 3 in the Cellos (mm. 191-199), imitating Violins at the distance of one measure, now creates this 10-measure section.

The next several sections also depart from an underlying 4-measure structure. The following 11-measure section (mm. 200-210, R. 12) groups into 2 + 5 (= 3 + 2) + 4. The overall organization of the following two sections is identical: both sections are 9 bars in length. But despite this similarity, the underlying organization of both these 9-bar sections differs: the first section (mm. 211-219) consists of two hypermeasures of 5 + 4 notated measures, while the following section consists of triple groupings of 3 + 3 + 3.[30]

The following sections return to hypermetric regularity, consisting of underlying four-bar units, including the 8-bar section at mm. 229-236, and the following 16-bar section at mm. 237-252. The return of the introductory material, now stated in tandem with Theme 3 (mm. 253-267) makes use of six statements of the ostinato motive α and a three-measure suffix.

The organization of the final 17 measures of the Allegro section (mm. 268-284) is intriguing in its reworking of Theme 1 material. Honegger reshuffles the contents of the original eight-measure theme, and reorders the material in two-measure units. We can show this by labeling the original eight-measure Theme 1 (from the exposition, mm. 30-37) in two-measure groupings of A (mm. 30-31), B (mm. 32-33), C (mm. 34-35), and D (mm. 36-37). In the final Allegro section, this material corresponds to: A (mm. 268-269), C (mm. 270-271), D (mm. 272-273), A (mm. 274-275), B (mm. 276-277), B (mm. 278-279), and final pitches of B (mm. 280-281).

Hypermetric Downbeats

In many cases, sectional divisions of the first movement are clear and evident. In other cases, however, Honegger takes pains to camouflage some of these divisions. He accomplishes this by using several techniques. First, he uses melodic material that ends on a hypermetric downbeat. This technique creates an overlap with the following section, making the hypermetric downbeat the goal of

[30] The final 3-measure grouping at mm. 220-222 consists of four measures of an expressed meter of 3/4, identical to its counterpart at mm. 94-96.

the preceding melodic material. Phrase slurs sometimes heighten this effect, and this technique appears in m. 54, m. 70 (Cellos), m. 117, m. 132 (Violins 1 and Cellos), m. 145, and m. 190. Often the subsequent melodic material begins after the hypermetric downbeat, forming an afterbeat pattern, such as at m. 54, m. 70, and m. 145.[31] In addition, melodic phrases at m. 117, m. 132, and m. 190 (Violins 1) begin squarely on the hypermetric downbeat, but these downbeats also form the endpoints of the previous phrases, resulting in phrase overlap on these hypermetric downbeats.[32] Honegger elides across the hypermetric downbeat most strongly at m. 85 (R. 5): rather than ending on the hypermetric downbeat, the previous phrase continues into the following beat. All these techniques serve to blur the hypermetric downbeats.

Elsewhere, Honegger masks some of the hypermetric downbeats by maintaining some of the motivic material from the previous section. The hypermetric downbeat at m. 127 is one potentially ambiguous spot. Here Violins 1 and Violas cease stating their Theme 1 material, yet Violas now pick up on the triplet β Motive, already stated in the previous section. As a result of this, it is certainly possible to consider m. 128, not m. 127, as the hypermetric downbeat.

Conclusion

Laurent Fichet suggests that an underlying hypermetric regularity consisting of 2, 4, and 8 measure phrases is characteristic of Honegger's rhythmic practice.[33] However, the preceding discussion shows that in the first movement of *Symphonie pour cordes* Honegger constructs large-scale sectional groupings out of hypermeasures of different and contrasting lengths. Only a few sections make use of an underlying four- and eight-bar hypermeter. These more regular sections appear when thematic material is first stated (Theme 1, mm. 30-54; Theme 3, mm. 70-77), the end of the exposition (mm. 97-104), as well as the measures that lead into the final return of the Introduction, mm. 229-236 and mm. 237-252. Elsewhere, the tight contrapuntal organization allows for an extreme flexibility in the lengths of sections and phrases, preventing an underlying hypermetric regularity.

---

[31] William Rothstein discusses afterbeat patterns in his book *Phrase Rhythm in Tonal Music* (New York and London: Schirmer Books, 1989), 29-30.

[32] For a discussion of phrase overlap, see Rothstein, *Phrase Rhythm*, 44-51; and Fred Lehrdahl and Ray Jackendoff, *A Generative Theory of Tonal Music* (Cambridge, Massachusetts: The MIT Press, 1983), 55-62.

[33] Fichet, "Problèmes de langage," *Honegger-Milhaud*, 118-19.

Chapter 6

Problems of Pitch Analysis: Centricity and Tonality

The central thesis of this book is that rhythmic and contrapuntal strategies provide a source of compositional unity to Honegger's output. Honegger disavowed the utility of precompositional tonal plans, but evidence such as the precompositional sketch for *Mouvement symphonique (Pacific 231)* (see Example 2.3) suggests that the composer used rhythmic precompositional plans, and such rhythmic plans may have *preceded* pitch organization.

This central thesis reverses the analytical assumption that pitch plays the fundamental role in determining musical structure, and that other elements, such as rhythmic organization, are secondary. While outside the primary focus of this book, the question of how best to address pitch organization in the music of Honegger cannot be ignored completely. Although the issue of pitch in Honegger's music remains an avenue for future inquiry, this chapter will summarize some problems, issues, and possible ways to proceed.

Pitch organization has provided the focal point for analytical approaches to other music written in the first half of the twentieth century, music coeval with that of Honegger. But despite a well-developed Schenkerian methodology for addressing tonal music of the Bach-to-Brahms tonal repertory, as well as a pitch-class set methodology for discussing structure in atonal music, the development of a successful analytical method has remained difficult for tonally centric music: music that is "organized around functional tone-centers but is not tonal in the traditional sense."[1]

Thus tonal centricity is an elusive topic, but it may be useful to identify how tonal centers can be employed on different levels of musical structure.[2] On

---

1 Joseph Straus, "A Principle of Voice Leading in the Music of Stravinsky," *Music Theory Spectrum* 4 (1982): 106. The terms tonal centricity, centricity, and neo-centricity are often used interchangeably.

2 For the tonally-centric music of Stravinsky, Joseph Straus offers some criteria for determining the central pitch class of a given passage. "First, a pitch class established in the lowest part and reinforced there by metric placement, reiteration, or duration tends to have priority over notes sounding above or around it. Second, in spite of the absence of traditional harmonic relations, 5/3 chords and seventh chords generally function as reinforcement for their lowest sounding pitch." See Straus's "Sonata Form in Stravinsky," in *Stravinsky Retrospectives*, ed. Ethan Haimo and Paul Johnson (Lincoln and London: University of Nebraska Press, 1987), 143.

the larger level, a composition may employ a global tonal center: a focal pitch class is prominent at the beginning and the ending of a composition or movement. Tonal centers may also be used to structure sections and phrases. On the smaller level, a tonal center may provide a musical arrival point within the context of a phrase.

To provide an introduction into some preliminary aspects of pitch organization in relation to Honegger's music, the chapter will critically examine some of Honegger's own statements on tonality, notation, and form in light of his actual compositional practice. The chapter will conclude with an overview of current analytical literature that addresses the tonally-centric music of Bartók, Stravinsky, and Hindemith.

Tonality *chez* Honegger?

Honegger's own writings seem dismissive of tonality. *Je suis compositeur* contains a number of statements that are critical of what he described as the "fetishism of tonality:"

> Gavoty: In this elaboration of your work, does the tonal design play a major role? For Vincent d'Indy, this plan played a determining role.
>
> Honegger: To me, the 'tonal design' is an abstraction. I do not even understand what it means. I have had numerous friendly debates with d'Indy himself on this topic. I told him at the time that to me the dogma of tonality seemed an entirely outmoded notion. As well require a contemporary dramatist to observe the unity of place of the classical tragedy, which so many masterpieces cheerfully ignore! ... Not everyone has 'absolute pitch.'
>
> Gavoty: It appears that Franck set down the tonal plan for his work at the top of his manuscript for his Symphony. It is said that he had decided upon it before having written a single note or jotted down the least theme.
>
> Honegger: It is a technique completely foreign to me, though I find it absolutely legitimate if the work benefits by it.[3]

Taking the composer's statements at face value, we must assume that tonality is not operative in the music. However, we should view Honegger's pronouncements on the "dogma of tonality" with caution. In one sense he is responding to d'Indy's compositional model, taken from Franck. D'Indy had described the Franck symphony as establishing an antagonism of tonalities in which two tonal poles are opposed until the ultimate dominance of one tonal pole over the other. In this sense the "tonal design" refers to a precompositional

---

[3] Honegger, *Composer*, 82-3.

sequence of keys, and this sequence creates the dialectic on which the work is founded.[4]

In another sense, perhaps we should take Honegger's statements as a refutation of diatonic or common-practice tonality. Indeed, rather than consistently atonal, much of the musical surface of his music is clearly organized around tonal centers, even in instances of near-aggregate saturation.[5]

However, what makes Honegger's remarks especially problematic is the fact that movements frequently begin and end around the same pitch class. This suggests that the composer still mandates an overarching tonal center for that movement, resulting in a large-scale tonal design determined by this beginning and ending tonal center.[6]

Moreover, on a larger level, constituent movements of many multi-movement works each end with the same tonal center. For example, all three movements of both the Second and the Fifth Symphony end with the pitch class or harmony based upon D. If the maintaining of a focal tonal center for a single movement seems reactionary, the practice of linking multi-movement works by the same tonal center is nearly atavistic by recalling Baroque practice. While in his writings Honegger rejects tonal design as an abstraction, this compositional practice conversely suggests tonal organization at its most global.

Even multi-movement works that do not maintain the same tonal center in all three movements nevertheless still exhibit a global design. For example, the outer movements of three-movement works such as the First Symphony, the Viola Sonata, and the Second String Quartet are linked by the same tonal center, while the central movements of each are tritone-related to their outer movements.

Notation and Tonality

At the very least, the composer's notational practice suggests a certain ambivalence toward tonality. As early as 1916 he began to abandon the use of key signatures. In two of the earliest song cycles, different songs appear both with and without key signatures. The *Trois poèmes de Paul Fort* (1916) contain two settings with key signatures, while the first is without a signature. In the *Six*

---

4 Vincent d'Indy, *Cours de composition musicale*, 3 vols. (Paris: Durand, 1902). This compositional model is discussed in further detail in Brian Jack Hart, "The Symphony in Theory and Practice in France, 1900-1914" (Ph.D. diss., Indiana University, 1994).

5 See Example 5.9 for an example of this. Some of the analytical literature that addresses issues of "tonal centricity" will be examined at the end of this chapter.

6 This may be due to Honegger's preoccupations with formal symmetry, and a fallout from his technique of reversed recapitulation. Similarly, Milton Babbitt suggests that archform symmetry within Bartók's String Quartets are used to generate closure in the absence of a clearly articulated tonality. See Milton Babbitt, "The String Quartets of Bartók," *Musical Quarterly* 35 (July 1949): 377-85.

poèmes d'Apollinaire (1916-1917), only the second and the sixth songs are given key signatures. Discussing a much later work, Kurt von Fischer reveals that the sketch for the slow movement of the Second String Quartet (1934-36) contains no key signature—only accidentals placed before each note—but that the published version, maintaining the identical sequence of pitches, contains a key signature of five sharps.[7] (Honegger did avow, however, that notational convenience was his reason for eliminating key signatures.)[8]

The composer's attitude toward orthography and his own statements in print also reflect an ambivalence toward tonality. In the 1940s he advocated the notational reforms proposed by Nikolas Obouhov, who had developed a system of notation that maintained the "white-note" piano pitches, but eliminated enharmonic duplication by representing chromatic alterations with a single symbol, a cross.[9] This championing of a notational system of only seven pitch-classes and their five chromatic alterations implies an indifference to tonality and tonal direction supplied by traditional enharmonic notation.[10]

Nevertheless, in Honegger's compositional practice, the composer is typically not indifferent to the role of notation in reflecting tonal direction and voice-leading. In fact, he is often quite careful to reveal tonal direction through his choice of accidentals. This is especially true in the case of ic1 (semitonal) voice-leading. The opening theme of the first movement of *Symphonie pour cordes* (mm. 30-37) offers a case in point. Example 6.1 contains the theme.

---

7 Multiple key signatures are rare in Honegger's music, although the opening of the third movement of the *Symphonie pour cordes* contains a key signature of C# major for Violins 1; the other strings are not given a key signature but articulate a D major harmony. Milhaud had attempted to codify some of his own ideas of polytonality in his 1923 article "Polytonalité et Atonalité," but his sole example of Honegger refers to a passage from *Horace Victorieux*, which he labels as atonal. The article appears in *La Revue Musicale* 4/4 (February 1, 1923): 29-44.

8 Arthur Honegger, *I Am a Composer*, trans. by Wilson O. Clough in collaboration with Allan Arthur Willman (New York: St. Martin's Press, 1966), 52-3. Originally published as *Je suis compositeur* (Paris: Editions du Conquistador, 1951).

9 Honegger became embroiled in a controversy in the press during the 1940s for his advocation of Obouhov's system, and he wrote the preface to Obouhov's *Traité d'Harmonie tonale, atonale et totale* (Paris: Durand, 1947). See Arthur Honegger, *Ecrits*, textes réunis et annotés par Huguette Calmel (Paris: Librairie Honoré Champion, 1992), 219, 237, 347-8, 490, 495-6. Honegger also wrote the *Deux esquisses* for piano (1943-1944) using Obouhov's notation.

10 The use of a composer's enharmonic notation as a source for understanding tonal direction in twentieth-century music is discussed in George Perle's article "Scriabin's Self-Analyses," contained in *The Right Notes* (Stuyvesant, NY: Pendragon Press, 1995), 3-26; and Malcom Gillies, "Pitch Notations and Tonality: Bartók," in *Early Twentieth-Century Music*, ed. Jonathan Dunsby (Oxford: Blackwell Publishers, 1993), 42-55.

Example 6.1. Measures 30-38, *Symphonie pour cordes*

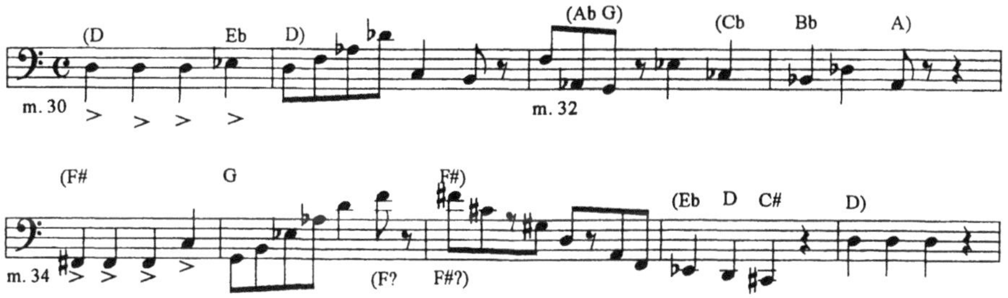

In almost every instance of adjacent or non-adjacent ic1 voice leading, Honegger takes pains to notate this as a stepwise connection. For example, in the first measure, the ic1 embellishment of D (the tonal center of the theme) is effected through Eb, not D#. Similarly, in m. 31 the final three pitches (Db, C, and B) reflect a stepwise relationship. Here this takes place even with the octave displacement between Db and C.

In the second and third pitches of m. 32, Honegger notates an Ab, which then moves to the G, again relying on a stepwise notation for the ic1 connection. At the very end of m. 32, Honegger uses Cb (rather than B) in order to reflect the motion to Bb at the downbeat of m. 33. The stepwise notation here also extends to the A on the third beat of the measure: unlike the Cb-Bb adjacency, the stepwise connection between Bb and A in m. 33 is nonadjacent, since the pitch Db intervenes in the second beat of this measure.

The analysis in Chapter Five called attention to the higher level connection F#-G-F#. These nonadjacent pitches occur on the downbeats of mm. 34, 35, and 36, and form a loose motivic association with the adjacent pitches D-Eb-D that occur from m. 30 to the downbeat of m. 31. The use of F# (rather than Gb) at mm. 34 and 36 shows again a stepwise connection to the G at m. 35.

It is interesting that this nonadjacent stepwise connection of F#-G-F# seems to trump a smaller-level ic1 voice-leading maneuver. The F# at the downbeat of m. 36 is directly preceded by F. In other instances, it would be likely that Honegger would notate the F# of m. 36 as Gb. However, it seems that the larger-level F#-G-F# connection, operating at the downbeats of mm. 34-36, motivates the choice of F#, rather than Gb, for the downbeat of m. 36. On the other hand, Honegger could easily have notated the final pitch of m. 35 as E#, rather than F, thus revealing a stepwise relation to the following F#. We can only speculate why Honegger did not.[11]

The final three pitches of the theme at m. 37 show the stepwise decoration of D (the tonal center of the theme) with both Eb and C#, and the notation works in support of the tonal direction.

Thus, despite Honegger's advocacy of Obouhov's notational system of only twelve pitches with no enharmonic duplication, in practice Honegger seems consistent and careful in his choice of notation, and prefers stepwise notation to reflect and clarify ic1 voice leading. Obviously, Honegger's pitch choices are not limited by the confines of diatonic or traditional tonal voice-leading, but he still seems to prefer to choose accidentals in order to reflect voice-leading in a roughly analogous fashion.

Form and Tonality: The Three String Quartets

One fruitful mode of inquiry for Honegger's pitch organization is the relationship of pitch to musical form. He spoke frequently of the importance of musical architecture, particularly in relation to sonata form. Using a metaphor from architecture, he advocated formal symmetry in a discussion of sonata form. He wrote:

> [T]he re-exposition of the order *A followed by B* is illogical, showing as it does, a defect in symmetry. In architecture you have a façade, the two left pillars of which, for instance, are arranged so that *A is followed by B*. Your object is solidly and harmoniously to rest the architrave and the pediment upon these pillars and upon those that form a pendant to them on the right. Symmetry as regards those on the right, will inevitably give you *B followed by A*, not vice versa, as the first time. Very well; in the sonata the pillars correspond to the themes; the pediment is the central development. Symmetrically, then, you must have *B followed by A* in the recapitulation since you had *A followed by B* in the exposition.

Honegger does not address the differences between the spatial symmetry of architecture and the temporal symmetry of music. Nevertheless, we can see that these ideas of formal symmetry are operative within all of the movements for string quartet written in sonata form.

He wrote three string quartets (1917, 1936, and 1937), each consisting of three movements. All three movements of the First String Quartet are in sonata

---

[11] Perhaps Honegger considered the F (following D) as a feature of the arpeggiation within m. 35—its notation as E#, while clarifying its relation to the following F#, would create an augmented second relation (D-E#) rather than its notated minor third relation. The notated minor third relation (D-F) seems to reveal its status as part of the m. 35 arpeggiation.

form, as are the outer two movements of both the Second and Third Quartets. In each of these sonata form movements, the recapitulation return of the complete statement of the opening theme is suppressed until the end of the movement, a technique commonly referred to as "reversed" or "reverse" recapitulation.[12]

Honegger's architectural metaphor quoted above suggests simply a five-part symmetry: the two pillars correspond to the first and second themes, and the central pediment corresponds to the development section. Taken at face value, this suggests an archform symmetry of ABCBA. In actual practice, Honegger's treatment of form and theme is somewhat more complicated.

Figure 6.1 provides a formal overview of the string quartet movements written in sonata form. The example shows the appearance of the themes in the exposition, and their reordering in the recapitulation. None of them exhibit simply a five-part archform symmetry, but all of the movements adhere to the principle of reversed recapitulation, in which a complete statement of the opening theme (1a) returns only at the end of the movement.[13] In all of these movements, much of the dramatic structure arises from the return of the principal idea at the end of the movement.

In several of the movements, we hear motivic fragments of the first theme intrude prior to the final restatement. This takes place, for example, in the First Quartet, first movement (m. 169), in the First Quartet, third movement (m. 196), and in the Second Quartet, first movement (m. 102). Nevertheless, a complete return of the opening theme occurs only at the end of the recapitulation.

Figure 6.1: Reversed recapitulation in sonata form movements of Honegger's String Quartets

First Quartet (1917)

First Movement

Exposition							
Theme:	1a	1b	2a	1b	2b	2c	Closing (from 1a)
Measure:	1	14	44	58	63	67	76
Reversed Recapitulation							
Theme:	2a	2b	2c	Transition		***1a**	(Coda)
Measure:	142	150	159	169		**189**	201

[12] Charles Rosen discusses "reverse" recapitulation in relation to Mozart's Sonata for Piano, K. 311 in *Sonata Forms* (New York and London: W.W. Norton and Company, 1988), 323.

[13] In fact, Honegger made revisions to the first and third movements of his First Quartet in order to accomplish reversed recapitulation. See Harry Halbreich, *Arthur Honegger*, trans. Roger Nichols (Portland, Oregon: Amadeus Press, 1992), 250.

Second Movement

Exposition

Theme:	Introduction	1a	1b	2a	2b
Measure:	1	15	26	41	65

Reversed Recapitulation

Theme:	1b	2a	2b	***1a**	Introduction
Measure:	96	111	125	**129**	142

Third Movement

Exposition

Theme:	1a	1b	2a	2b	2c	2d (Closing)
Measure:	1	26	33	55	74	87

Reversed Recapitulation

Theme:	2a	2b	New	Fragments of 1a	2c	2d	***1a**	1a + 2a (Coda)
Measure:	140	170	187	196	200	213	**222**	252

Second String Quartet (1936)

First Movement

Exposition

Theme:	1a	1b	1a	1b	2a	2b	2c	2d
Measure:	1	19	23	34	36	43	46	56

Reversed Recapitulation

Theme:	2b (with fragments of 1a)	2c + 2a	2d (developed)	1b (dev.)	***1a**
Measure:	102	110	123	143	**149**

Third Movement

Exposition

Theme:	1a	1b	2a	1a
Measure:	1	24	32	47

Recapitulation

Theme:	1b	2a + ***1a**
Measure:	103	**114**

Third String Quartet (1937)

First Movement

Exposition

Theme:	1	1 (developed)	2a	2b
Measure:	1	21	31	41

Reversed (and developing) Recapitulation

Theme:	2a (developed)	2b + 2a (developed)	***1a**	Coda (Th. 1)
Measure:	108	120	**140**	156

Third Movement

Exposition

Theme:	1a	1b	2a	2b
Measure:	1	31	40	60

Reversed (and developing) Recapitulation

Theme:	1b	2a	***1a** + 1b + 2a + 2b	***1a** (augmentation canon)
Measure:	128	134	**162**	**179**

In a quotation of Honegger's examined earlier in the chapter, he dismissed D'Indy's notion of the "tonal plan" as an abstraction. However, in a number of these quartet movements, Honegger seems to rely on a systematic plan in order to play into and dramatize the reversed recapitulation. This plan does rely on the overall pitch design of these movements.

Honegger is most systematic in his First String Quartet. Figure 6.2 begins with the formal organization of the recapitulations of all three movements of the First Quartet. In addition, the figure provides the transpositional level of the recapitulation statement of the themes relative to the statement of that theme in the exposition. "T" refers to transpositional level: Therefore T0 refers to a recapitulation thematic statement at the same pitch level as at the exposition, T1 refers to a recapitulation thematic statement a half step above the exposition statement of the theme, T-1 refers to a half step below the exposition statement of the theme, etc.

In the First Quartet, first movement, the recapitulation begins with Theme 2a, stated a minor third (T-3) below the level of the exposition. Theme 2b is stated first a tritone, then a minor third below the exposition statement, and 2c is stated a minor third below the exposition statement. The return of the principal 1a theme coincides with a statement at the original transpositional level (T0) of the theme. Thus, the overall thrust of the reversed recapitulation seems to be the ultimate statement of Theme 1a that occurs at the identical transpositional level as at the beginning of the composition.

In the second and third movements from the First String Quartet, Honegger is even more systematic in bringing about the final statement of Theme 1a. In the second movement, the recapitulation statement of Theme 1b enters a whole step above (T2) the exposition statement. The following Themes 2a and 2b appear a half step above (T1) the exposition statement. As in the first movement, the final statement of Theme 1a occurs at the original pitch level of T0. In this movement, notice the systematic transpositional scheme in the recapitulation of T2—T1—T0: with each successive thematic statement, the transpositional levels move closer to the final T0 statement of Theme 1a.[14]

Honegger uses a similar scheme in the final movement. The recapitulation restatement of 2a appears both at T3 and T4. Theme 2b appears at T4, and Themes 2c and 2d appear at T2. Now, however, at m. 222 a statement of Theme 1a appears at T1, a half step above the original exposition statement. The transpositional appearance of the principal theme a half step above its earlier statement creates a tonal argument. This brings about the dramatic high point of the composition at mm. 237-251, and at m. 246 a fragment of Theme 1a is voiced in a highly dissonant seven-pitch chord, stating the pitch-class set <013468T>. Following, Theme 1a is finally stated back at the original pitch level (now C major rather than C minor, however) in the coda.

In all three movements of the string quartet, the reversed recapitulation is effected not only through the suppression of the opening theme of the exposition until the end of the recapitulation, but also through a return to the original pitch level of that theme. In addition, in the second movement Honegger systematically states the previous themes at systematically smaller transpositional levels (T3—T2—T1) until the final T0 statement of Theme 1a. Toward the end of the third movement, Honegger dramatizes a T1 statement of Theme 1a, until it is finally stated at the original T0 level in the coda.[15]

---

[14] Honegger's statements on reversed recapitulation acknowledge the notion of symmetry for providing the motivation for reversed recapitulation. This movement also contains an additional level of symmetry, since the introduction frames the beginning and ending of the movement.

[15] In the other sonata form movements of String Quartets 2 and 3, Honegger similarly restates the principal theme at the end of the recapitulation at the same transpositional level as the exposition. In some movements it is only at the final measures that the T0 transpositional relation is made clear: for example, only the final seven measures of the first movement of Honegger's third String Quartet makes use of the G-E melodic cell that opens the movement. In general, compared to the First Quartet, he is less systematic about the transpositional plans for the other thematic statements that occur prior to the final T0 recapitulation statement of Theme 1a. In the first movement to the Second String Quartet, however, Themes 2c and 2a are stated in the recapitulation (m. 110) at T7 and T-1 relative to the exposition, while Theme 2d (m. 123) occurs at T1. Very possibly, Honegger was interested in exploring these T-1 and T1 relations, symmetrical around the T0 at the final statement of 1a at m. 149.

Figure 6.2: Transpositional schemes in the reversed recapitulations in Honegger's First Quartet, relative to the exposition (T1 = Recapitulation statement of theme is half step above exposition statement of theme, T-1 = Recapitulation statement of theme is half step below exposition statement of theme, etc.)

First Quartet (1917)

First Movement
Reversed Recapitulation

Theme:	2a	2b	2c	Transition	***1a**	(Coda)
Measure:	142	150	159	169	**189**	201
Transposition:	T-3	T6,T-3	T-3		**T0**	

Second Movement
Reversed Recapitulation

Theme:	1b	2a	2b	***1a**	Introduction
Measure:	96	111	125	**129**	142
Transposition:	T2	T1	T1	**T0**	T0

Third Movement
Reversed Recapitulation

Theme:	2a	2b	1a: Fragments	2c	2d	***1a**	**1a** + 2a
Measure:	140	170	196	200	213	**222**	**252**
Transposition:	T3,T4	T4		T2	T2	**T1**	**T0**

Tonal Centricity: Analytical Approaches

This final section examines some analytical approaches for the tonally-centric music of Bartók, Stravinsky, and Hindemith. Since these inquiries are often composer-specific and their analytical methods do not always overlap, this calls into question the utility and the possibility of developing a synthesis or a general theory for tonally-centric music. Points of intersection rest upon two general questions:

> 1.) *How are tonal centers articulated?*
> 2.) *Can we understand some level of cooperation between different centers, or do they operate relatively independently of one another?*

In answer to the first query, Elliott Antokoletz suggests that a tonal center may be established in one of two ways in the music of Bartók: 1) a pitch class may be the primary or focal tone of a traditional mode; or 2) centricity can be established by an axis of inversion given or implied by pitch and pitch-class symmetry. In an analysis of Bartók's First String Quartet, Antokoletz argues that "a sense of tonal priority is apparent within every textural fiber and at every cadential turn in this quartet and, indeed, in all of Bartók's compositions."[16]

In Bartók's First String Quartet, the vacillation between A major and A minor in the First Quartet fulfills Antokoletz's modal condition for establishing tonal priority. However, Antokoletz almost immediately leaves behind both his modal or axial conditions, observing that tonal priority in the Quartet is also articulated by "nontraditional configurations that are built upon the pitch-class A."[17] Thus in the absence of modality or inversional symmetry, tonal weight is assigned to the lowest pitch of non-triadic sonorities. Here Antokoletz finesses one of the thornier issues of deducing tonal priority from harmonies that are neither triadic nor symmetrical, conferring priority upon the acoustically lowest-sounding pitch.

In a chapter entitled "Tonality," Daniel Parker enumerates factors that contribute to or obscure what he refers to as "tonal focus" in the symphonies of Honegger.[18] He notes that "tonality is deemed present in a phrase (or larger formal unit) when prominence as a focal or referential point is assumed by a pitch (or pitches) such that other pitches are dependently, though perhaps not hierarchically, related to it."[19] Elements that contribute to tonal focus include melodic activity, pitch reiteration, pedal, cadential material, and combinations of these factors. For Parker these elements are sometimes reinforced by tertian harmony and metric placement. On the other hand, Parker enumerates conditions that obscure tonal focus: tonally ambiguous harmonic structures and backgrounds, tonally ambiguous layering of harmonic structures and backgrounds, tonally ambiguous melodic constructs, and tonally ambiguous combinations of melodic and harmonic materials. Parker's data reveals that Honegger's Fourth Symphony contains the highest percentage of measures that contain a clear tonic (92%), while the Fifth Symphony contains the smallest percentage (38%).

Parker's criteria for determining tonal focus addresses our first question above, that of how tonal centers (or "tonics") are articulated. However, Parker's data does not address our second question, that of the coordination of different centers within a composition. Parker merely suggests the presence of multiple local "tonics" within movements. Absent too is the notion of a larger tonal context upon which these local "tonics" are dependent: are movements controlled by a single "tonic," key, pitch, or pitch class? The statistical results do not

16 Elliott Antokoletz, *The Music of Béla Bartók: A Study of Tonality and Progression in Twentieth-Century Music* (Berkeley: University of California Press, 1984), 142. See especially the chapter "Tonal Centricity Based on Axes of Symmetry," pp. 138-203.

17 Ibid.

18 Daniel Quienton Parker III, "The Symphonies of Arthur Honegger: An Examination of Stylistic Factors and Form" (Ph.D. dissertation., University of Kentucky, 1990), 139-86.

19 Ibid., 140.

engage the question.

The following examines here two ways in which the analytical literature treats the relationship of different tonal centers to one another in a composition. One method retains the notion of harmonic function from common-practice tonality. This approach purports a controlling tonic upon which the other harmonic functions are dependent. A second method adapts Schenkerian reductive methods. The latter attempts to show how surface design motivates larger-level arrival points.

Harmonic Function

The three harmonic functions of tonic, dominant and subdominant serve to articulate a tonality. Traditionally the dominant and subdominant harmonies are symmetrically positioned around the tonic triad, a perfect fifth above and below. Chords not specifically tonic, subdominant, and dominant may nevertheless be collapsed into one (or more) of these three functional categories. Function is explanatory of foreground harmonic—especially cadential—progression. In addition, harmonic function defines larger interior key regions by their relation to the tonic, and incorporates them into one of the three functional categories.

Ernő Lendvai's reorientation of the functional categories brings *each* chromatic degree, relative to tonic, into the sphere of the three harmonic functions.[20] For the music of Bartók, he posits that functional equivalence is determined by inclusion within a minor third cycle. Thus, given a tonic key of a composition, keys a minor third and tritone away from that tonic also fulfill tonic function. Lendvai refines the three functional categories further by distinguishing between a principal branch (the original key and its tritone partner) and the secondary branch (keys three semitones above and below the tonic). The subdominant function appears upon the subdominant (IV) and its minor third cycle partners, while the dominant function appears upon the dominant (V) and the other keys a minor third and tritone apart.

Paul Wilson expands Lendvai's ideas of harmonic function in the music of Bartók.[21] Instead of determining function through an intervallic or positional relationship to the tonic, Wilson determines function through an examination of relative contextual relationships. Function is thus considered or defined by its action or activity. He extends the list to five functions, including tonic,

---

[20] Ernő Lendvai, *Béla Bartók: An Analysis of His Music* (London: Kahn and Averill, 1971), and *The Workshop of Bartók and Kodály* (Budapest: Editio Musica, 1983). Arnold Whittall presents a detailed analysis of Bartók's Second String Quartet based upon Lendvai's axis system. He argues that the piece "presents one of Bartók's most rigorously organized arguments between tonal centers" in his book *Music Since the First World War* (New York: St. Martin's Press, 1977), 33-41.

[21] See Paul Wilson, *The Music of Béla Bartók* (New Haven and London: Yale University Press, 1992), especially 32-9.

dominant, subdominant, dominant preparation, and tonic substitution or extension. Tonic function is then allied with the principal goal tone, determined or corroborated by non-pitch elements such as form, rhythm, contour, dynamics, or other musical elements. Wilson suggests two types of dominant: 1) a "local dominant" that prepares and is followed by the arrival of the goal tone, and 2) a large-scale dominant that is an interior tonal center or goal event within a composition.[22]

Neumeyer's study of Hindemith relies on the notion of intervallic position in order to determine harmonic function. Function here extends to each scale degree of the key.[23] Neumeyer's theory is rooted in Hindemith's theoretical and compositional writings. As a result, he determines the strength or force of harmonic progression by Hindemith's Series I, which places in hierarchy the relationship of a single pitch to all chromatic degrees.[24] Strength of progression is then determined by proximity to the focal pitch along the series. As in tonal music, function is potentially relative and moveable, and may be based upon local harmonic context. For example, harmonic motion down a perfect fifth may embody the same "strong" harmonic function even when the destination is not the original key of the composition.

Reductive Methodologies

As the brief summary above suggests, harmonic function in twentieth-century music explains that the tonal centers of a composition represent one of the tonal functions. On the other hand, reductive methods may also explain the sequence of different tonal centers, but analysts often explain this sequence by correlating surface musical design with longer-range and structurally-weighted arrival points. This requires a modified Schenkerian orientation. Retained from a Schenkerian methodology are correlations of "foreground" to "middleground." Discarded is a predetermined or absolute notion of tonality articulated through contrapuntal prolongation and elaboration of a background tonic-dominant axis. These modified Schenkerian approaches often suggest that composers deploy in the linear dimension the pitch-class sets that are prominent on the musical surface.

For example, an article by Joseph Straus on Stravinsky corresponds pervasive features on the musical surface with longer-range arrival points.[25] Here

---

22 By this criterion (and through interpretation of Bartók's prose writing), Wilson then posits Bartók's tritone relationship—a frequent harmonic destination within the interior of a movement—as embodying a dominant function. This contrasts with Lendvai's labeling of this harmonic area as tonic. See Wilson, *Bartók*, 37.

23 David Neumeyer, *The Music of Paul Hindemith* (New Haven and London: Yale University Press, 1986).

24 Series I is presented in Paul Hindemith, *The Craft of Musical Composition, Vol. 1: Theoretical Part*, trans. Arthur Mendel (New York: Associated Music, 1942, rev. ed., 1945).

25 Straus, "A Principle of Voice-Leading," 106-24. Foundations for this approach may

an unordered collection of pitch classes prevalent at the surface of the music—a referential collection—governs longer-range motion. Straus asserts that the linear completion of this collection offers long-range directed motion. Typically the arrival points that compose out the collection are prominent pitches in the bass.

Similarly, Wilson's study of Bartók relaxes the Schenkerian voice-leading model yet engages Schenkerian reductive methodology and notions of hierarchical structure in order to illuminate large-scale movement within the compositions.[26] Context-sensitive criteria contribute to hierarchical weight. These criteria include: 1) recurrent transpositional intervals ("privileged patterns"), 2) pitch-class sets on the musical surface that return at temporal boundaries or structural positions ("projected sets"), 3) departure and return, and 4) functional categories.

At the beginning of this chapter, two fundamental questions were posed: how are tonal centers articulated, and how do they correspond to one another. Straus and Wilson both address our first question by providing criteria to determine tonal centers: "hierarchical weight" for Wilson, and prominent arrival or cadential points for Straus. Both authors also address the second question by explaining the sequence of tonal centers as hierarchically-weighted points that project events from the musical surface. In this sense, the analyses reveal the overall directed motion of the compositions under investigation. Yet the analytical claims are more modest than traditional Schenkerian analyses that attempt to demonstrate organic unity through prolongational structures.[27] In addition, we might note that Wilson's and Straus's analyses are piece-specific and contextual, and their modified Schenkerian approaches do not necessarily require the presence of a single governing tonal center for an entire composition.

---

be detected in Allen Forte, "New Approaches to the Linear Analysis of Music," *Journal of the American Musicological Society* 41/2 (1988): 315-48. See also the Straus article "Stravinsky's Tonal Axis," *Journal of Music Theory* 26/2 (1982): 261-90, in which Straus shows large-scale harmonic motion within the course of a composition to be the motion from one major or minor triad to another "interlocking" major or minor triad which possesses two pitches in common with the original triad. The criterion for determination of this triad is based upon its use as a "referential sonority, especially at cadences."

[26] Paul Wilson, *Bartók*.

[27] Prolongational structures in twentieth-century music provided the basis for some earlier analytical forays, especially those by Felix Salzer and Roy Travis. For a brief history and refutation of these Schenkerian enterprises, see Joseph Straus, "The Problem of Prolongation in Post-Tonal Music," *Journal of Music Theory* 31 (1987): 1-21. See also James Baker, "Schenkerian Analysis and Post-Tonal Music," in *Aspects of Schenkerian Theory*, ed. by David Beach, 153-186; and William Rothstein, "The Americanization of Heinrich Schenker," *In Theory Only* 9/1 (March 1986): 5-17.

Coda

The analytical approaches summarized above have been faintly echoed during the course of this book. As with Antokoletz' two types of tonal centricity, we have seen that tonal centers are established in both manners in the music of Honegger. Tonal centers are established in the presence of the diatonic collection (Example 1.5), as well as through instances of pitch symmetry (Examples 3.7 and 3.8). In addition, as in the discussion of reductive methodologies of Wilson and Straus in this chapter, the analysis of *Rugby* describes the projection of surface elements—the C#-D leading-tone motion that opens the composition—into the sequence of harmonic areas that closes the first half of the composition (Figure 4.4).

The analyses in this book typically do not discuss the notion of a larger-level tonality or centricity that extends beyond these interior tonal centers. This chapter is intended merely to highlight some possibilities for future examinations of the role of pitch organization in the music of Honegger, considering the role of tonality, centricity, notation, and form. In addition, the issue of tonal centricity in Honegger's music can very possibly build upon some of the analytical inroads made for the music of Stravinsky, Bartók, and Hindemith.

Another unresolved issue of pitch is that several of Honegger's works, such as *Horace Victorieux* (1921) and the opera *Antigone* (1927), are atonal works, lacking the presence of large-scale or interior tonal centers. These works require an altogether different analytical tack. Certainly techniques of inversional symmetry (as discussed in Chapter 3) govern some of the musical surface, but significant portions of these works are written in a free atonal orientation, suggesting a pitch-class set methodology for analysis. Although perhaps less characteristic of the composer, many of his younger contemporaries deemed these works the most potent of Honegger's, and we may again recall Pierre Boulez:

> *Antigone*, *Horace Victorieux*, the *Symphonie pour cordes*: it is from there that we began a journey rich in surprises. Let us thank Honegger for having given us the taste for adventure.[28]

---

[28] "*Antigone*, *Horace Victorieux*, la *Symphonie pour cordes*, c'est par là que nous avons commencé un périple fertile en surprises. Qu'Honegger soit remercié de nous avoir donné le goût de l'aventure." Pierre Boulez, Obituary from "Hommage à Honegger," *Melos* (January 1956): 4.

Bibliography

Antokoletz, Elliott. "Principles of Pitch Organization in Bartók's Fourth String Quartet." *In Theory Only* 3/6 (September 1977): 3-22.

—. *The Music of Béla Bartók: A Study of Tonality and Progression in Twentieth-Century Music*. Berkeley and Los Angeles: University of California Press, 1984.

—. "Interval Cycles in Stravinsky's Early Ballets." *Journal of the American Musicological Society* 39/3 (Fall 1986): 578-614.

Archibald, Bruce. "Some Thoughts on Symmetry in Early Webern: Op. 5, No. 2." *Perspectives of New Music* 10/2 (Spring/Summer 1972): 159-63.

Babbitt, Milton. "The String Quartets of Bartók." *Musical Quarterly* 35 (July 1949): 377-85.

Baker, James. "Schenkerian Analysis and Post-tonal Music." In *Aspects of Schenkerian Theory*, ed. David Beach. New Haven and London: Yale University Press, 1983.

—. *The Music of Alexander Scriabin*. New Haven: Yale University Press, 1986.

Benjamin, William. "Ideas of Order in Motivic Music." *Music Theory Spectrum* 1 (1979): 23-4.

—. "A Theory of Musical Meter." *Music Perception* 1/4 (1984): 376-413.

Bernard, Jonathan. "Space and Symmetry in Bartók." *Journal of Music Theory* 30/2 (Fall 1986): 185-201.

—. *The Music of Edgar Varèse*. New Haven: Yale University Press, 1987.

Berry, Wallace. *Structural Functions in Music*. Englewood Cliffs: Prentice-Hall, 1976.

—. "Symmetrical Interval Sets and Derivative Pitch Materials in Bartók's String Quartet No. 3." *Perspectives of New Music* 18 (1979-1980): 287-379.

—. "Metric and Rhythmic Articulation in Music." Music Theory Spectrum 7 (1985): 7-33.

Boulez, Pierre. "Stravinsky Remains." In *Stocktakings of an Apprenticeship*, trans. Stephen Walsh. London: Clarendon Press, 1991.

Brody, Elaine. *Paris: The Musical Kaleidoscope 1870-1925*. New York: George Braziller, Inc., 1987.

Bruyr, José. *Arthur Honegger et son oeuvre*. Paris: Editions Corrêa, 1947.

Cohn, Richard L. "Inversional Symmetry and Transpositional Combination in Bartók." *Music Theory Spectrum* 10 (1988): 19-42.

—. "The Dramatization of Hypermetric Conflicts in the Scherzo of Beethoven's Ninth Symphony." *19th-Century Music* 15/3 (1992): 22-40.

—. "Metric and Hypermetric Dissonance in the *Menuetto* of Mozart's Symphony in G Minor, K. 550." *Intégral* 6 (1992): 1-33.

Collaer, Paul. *Antigone 1928 (Etude)*. Paris: Editions Senart, 1928.

Delannoy, Marcel. *Honegger*. Paris: Pierre Horay, 1953. Nouvelle édition augmentée du catalogue des oeuvres de A. Honegger par G.K. Spratt. Geneva: Editions Slatkine, 1986.

Dewey, Cindy Neumayer. "A Performer's Guide to the Songs of Arthur Honegger." DMA Thesis, The Louisiana State University and Agricultural and Mechanical College, 1994.

Feschotte, Jacques. *Arthur Honegger*. Paris: Editions Seghers, 1966.

Fischer, Kurt von. "Arthur Honegger." In *Essays in Musicology*, trans. Carl Skoggard, ed.

Tamara S. Evans. New York: The Graduate School and University Center, City University of New York, 1989.

Friedmann, Michael. "A Methodology for the Discussion of Contour: Its Application to Schoenberg's Music." *Journal of Music Theory* 29 (1985): 223-48.

Gédalge, André. *Treatise on the Fugue*. Trans. and ed. Ferdinand Davis. Norman, Oklahoma: University of Oklahoma Press, 1965.

George, André. *Arthur Honegger*. Paris: Editions Aveline, 1926.

Goldberg, Leslie Ann. "An Analysis of Harmonic Language in the Songs of Arthur Honegger." Mus.A.D. Thesis, Boston University, 1991.

Halbreich, Harry. *Arthur Honegger: Un musicien dans la cité des hommes*. Librairie Arthème Fayard, 1992.

—. *L'oeuvre d'Arthur Honegger*. Paris: Honoré Champion Éditeur, 1994.

—. *Arthur Honegger*, trans. Roger Nichols. Portland, Oregon: Amadeus Press, 1999.

Harrison, Daniel. *Harmonic Function in Chromatic Music*. Chicago: University of Chicago Press, 1994.

Hart, Brian Jack. "The Symphony in Theory and Practice in France, 1900-1914." Ph.D. Dissertation, Indiana University, 1994.

Headlam, David. *The Music of Alban Berg*. New Haven and Yale: Yale University Press, 1996.

Honegger, Arthur. *Incantation aux fossiles*. Lausanne: Editions d'Ouchy, 1948.

—. *I am a Composer*, trans. Wilson Clough. New York: St. Martin's Press, 1966. Originally published as *Je Suis Compositeur*. Paris: Editions du Conquistador, 1951.

—. *Nachklang, Schriften, Photos, Dokumente*, ed. Willi Reich. Zurich: Die Arche, 1957.

—. *Ecrits*, Textes réunis et annotés par Huguette Calmel. Paris: Librairie Honoré Champion, 1992.

Horlacher, Gretchen. "The Rhythms of Reiteration: Formal Development in Stravinsky's Ostinati." *Music Theory Spectrum* 13/1 (1991): 61-78.

Kaminsky, Peter. "Aspects of Harmony, Rhythm and Form in Schumann's *Papillons*, *Carnaval*, and *Davidsbündlertanze*. Ph.D. Dissertation, University of Rochester, 1989.

Kelkel, Manfred, ed. *Honegger-Milhaud: Musique et Esthétique*, Actes du Colloque International tenu à la Sorbonne en novembre 1992. Paris: Librairie J. Vrin, 1994.

Krebs, Harald. "Some Extensions of the Concept of Metrical Consonance and Dissonance." *Journal of Music Theory* 31/1 (1987): 99-120.

Lambert, Philip. "Interval Cycles as Compositional Resources in the Music of Charles Ives." *Music Theory Spectrum* 12/1 (1990): 43-83.

Landowski, Marcel. *Honegger*. "Solfèges" Series. Paris: Editions du Seuil, 1957.

Lendvai, Ernő. *Béla Bartók: An Analysis of His Music*. London: Kahn and Averill, 1971.

—. *The Workshop of Bartók and Kodály*. Budapest: Editio Musica, 1983.

Lerdahl, Fred and Ray Jackendoff. *A Generative Theory of Tonal Music*. Cambridge, Mass.: The MIT Press, 1983.

Lester, Joel. *The Rhythms of Tonal Music*. Carbondale and Edwardsville: Southern Illinois University Press, 1986.

Lewin. David. "Inversional Balance as an Organizing Force in Schoenberg's Music and Thought." *Perspectives of New Music* 6/2 (1968): 1-21.

—. "Some Investigations into Foreground Rhythmic and Metric Patterning." In *Music Theory: Special Topics*, ed. Richmond Browne. New York: Academic Press, 1981.

—. "Vocal Meter in Schoenberg's Atonal Music with a Note on Serial Hauptstimme." *In Theory Only* 6/4 (1982): 12-36.

—. "Transformational Techniques in Atonal and Other Music Theories." *Perspectives of New*

Music 21/1-2 (Fall/Winter 1982 and Spring/Summer 1983): 312-71.
—. *Generalized Musical Intervals and Transformation*. New Haven: Yale University Press, 1987.
McNamee, Ann. "Bitonality, Mode, and Interval in the Music of Karol Szymanowski." *Journal of Music Theory* 29/1 (1985): 61-84.
Maillard, Jacques and Jacques Nahoum. *Les Symphonies d'Arthur Honegger*. Paris: Alphonse Leduc, 1974.
Marvin, Elizabeth and Paul Laprade. "Relating Musical Contours: Extensions of a Theory for Contour." *Journal of Music Theory* 31 (1987): 225-67.
Mason, Colin. "An Essay in Analysis: Tonality, Symmetry, and Latent Serialism in Bartók's Fourth Quartet." *Music Review* 18 (1957): 189-201.
Matter, Jean. *Arthur Honegger ou la quête de joie*. Lausanne and Paris: Editions Foetisch, 1956.
Meylan, Pierre. *René Morax et Arthur Honegger au Théâtre du Jorat*. Paris: Ploix, 1966.
—. *Arthur Honegger: Humanitare Botschaft der Musik*. Frauenfeld and Stuttgart: Verlag Huber & Co., 1970.
Milhaud, Darius. *Notes without Music*. Translated by Donald Evans. New York: Alfred A. Knopf, Inc., 1953.
Morgan, Robert P. *Twentieth-Century Music: A History of Musical Style in Modern Europe and America*. New York and London: W.W. Norton & Co., 1991.
Morris, Robert. "New Directions in the Theory and Analysis of Musical Contour." *Music Theory Spectrum* 15/2 (Fall 1993): 205-28.
Neumeyer, David. *The Music of Paul Hindemith*. New Haven and London: Yale University Press, 1986.
Parker, Daniel Quienton II. "The Symphonies of Arthur Honegger: An Examination of Stylistic Factors and Form." Ph.D. Dissertation, University of Kentucky, 1990.
Perle, George. "Symmetrical Formations in the String Quartets of Béla Bartók." *Music Review* 16 (1955): 300-312.
—. Berg's Master Array of the Interval Cycles." *Musical Quarterly* 63/1 (1977): 1-30.
—. *Twelve-Tone Tonality*. Berkeley and Los Angeles: University of California Press, 1977.
—. *The Right Notes*. Stuyvesant, New York: Pendragon Press, 1995.
Perloff, Nancy. *Art and the Everyday: Popular Entertainment and the Circle of Erik Satie*. Oxford: Clarendon Press, 1991.
Porter, Charles. "Interval Cycles in Alban Berg's String Quartet Opus 3." *Theory and Practice* 14/15 (1989-1990): 139-78.
Poulenc, Francis. *Moi et mes amis. Confidences recueillies par Stéphane Audel*. Geneva and Paris: La Palatine, 1963.
Roland-Manuel. *Arthur Honegger*. Paris: Editions Senart, 1925.
Rostand, Claude. *La Musique française contemporaine*. Paris: Presses Universitaires de France, 1952.
Roy, Jean. *Le groupe des six*. Paris: Editions Seuil, 1994.
Salzmann, Eric. *Twentieth-Century Music: An Introduction*. 2nd edn., Englewood Cliffs, New Jersey: Prentice-Hall Inc., 1980.
Schachter, Carl. "Rhythm and Linear Analysis: Aspects of Meter." *The Music Forum* 6 (1987): 1-59.
Shadel, Mary Louise. "Thematic Materials and Developmental Procedures in the Slow Movements of the Five Symphonies of Arthur Honegger." Master's Thesis, Georgia State University, 1982.
Shattuck, Roger. *The Banquet Years: The Origins of the Avant-Garde in France 1885 to World War I*. Revised edition. New York: Random House, 1968.

Spratt, Geoffrey K. *The Music of Arthur Honegger*. Cork, Ireland: Cork University Press, 1987.
Straus, Joseph. "A Principle of Voice Leading in the Music of Stravinsky." *Music Theory Spectrum* 4 (1982): 106-24.
—. "Stravinsky's Tonal Axis." *Journal of Music Theory* 26/2 (1982): 261-90.
—. "The Problem of Prolongation in Post-Tonal Music." *Journal of Music Theory* 31 (1987): 1-21.
Sysoeva, Elena. *Simfonii A. Honeggera*. Moscow: Muzyka, 1975.
Tappolet, Willy. *Arthur Honegger*. Translated into French by Claude Tappolet. Neuchâtel: Editions La Baconnière, 1957.
Treitler, Leo. "Harmonic Procedures in the Fourth Quartet of Béla Bartók." *Journal of Music Theory* 3 (1959): 292-98.
van den Toorn, Pieter. "Rhythmic or Metric Invention." In *The Music of Igor Stravinsky*. New Haven: Yale University Press, 1983.
Watkins, Glen. *Soundings: Music in the Twentieth Century*. New York: Schirmer Books, 1988.
—. *Pyramids at the Louvre: Music, Culture, and Collage from Stravinsky to the Postmodernists*. Cambridge, Massachusetts and London: The Belknap Press of Harvard University Press, 1994.
Wilson, Paul. *The Music of Béla Bartók*. New Haven and London: Yale University Press, 1992.
Yeston, Maury. *The Stratification of Musical Rhythm*. New Haven: Yale University Press, 1976.

Index

I

J

K

L

M

N

O

P

R

S

For Product Safety Concerns and Information please contact our EU representative GPSR@taylorandfrancis.com
Taylor & Francis Verlag GmbH, Kaufingerstraße 24, 80331 München, Germany

www.ingramcontent.com/pod-product-compliance
Ingram Content Group UK Ltd.
Pitfield, Milton Keynes, MK11 3LW, UK
UKHW022025061126
477531UK00002B/45
9781138728691